Concurrent Engineering

What's working where

The Design Council

The Design Council is the UK's national authority on design. Its main activities are:

- commissioning research on design-related topics, particularly stressing design effectiveness to improve competitiveness;

- communicating key design effectiveness messages to industry and government; and

- working to improve both design education and the role of design in education generally.

The Design Council is working with Gower to support the publication of work in design management and product development. For more information about the Design Council please phone 0171-208 2121. A complete list of book titles is available from Gower Publishing on 01252 331551.

Concurrent Engineering

What's working where

Edited by
Dr Christopher J Backhouse
and
Dr Naomi J Brookes

The Design Council

Gower

Published by
Gower Publishing Limited
Gower House
Croft Road
Aldershot
Hampshire GU11 3HR
England

Gower
Old Post Road
Brookfield
Vermont 05036
USA

British Library Cataloguing in Publication Data
Concurrent engineering: What's working where
  1. Concurrent engineering
  I. Backhouse, Christopher J. II. Brookes, Naomi J. III. Design
Council
620'.0042

ISBN 0 566 07666 7

Library of Congress Cataloging-in-Publication Data
Concurrent engineering: What's working where / edited by Christopher J.
  Backhouse and Naomi J. Brookes.
     p. cm.
   ISBN 0-566-07666-7 (hardcover)
    1. Production engineering. 2. Concurrent engineering.
  I. Backhouse, Christopher J. II. Brookes, Naomi J.
  TS155.C595   1996
  658.5—dc20   96-4825
                                           CIP

Typeset in New Baskerville by Bournemouth Colour Press, Parkstone and printed in Great Britain by Hartnolls Ltd, Bodmin.

Contents

List of Figures and Tables

Figures

Tables

Preface

Concurrent engineering has been accepted as the most influential initiative affecting new product introduction in this, the last decade of the twentieth century. As global markets develop to demand increasing numbers of higher quality products at lower cost, pressures on companies to supply suitable products are growing ever more intense. As technology advances to facilitate ever more product features, life-cycles are diminishing and products are reaching obsolescence ever more quickly. Previous experience of coping with this situation through extending product life-cycles is no longer valid: the competition is so fierce that new products are arriving all the time. The initiative of concurrent engineering is therefore aimed at maximizing profitability through the shortening of new product introduction time.

Although the overall philosophy of concurrent engineering and its aims can be assumed to be well understood, the precise details of what form of concurrent engineering is most appropriate for any given set of circumstances are not so well defined. Specific cases of successful implementation do not necessarily ensure success if repeated elsewhere. Whilst all companies are experiencing pressure to shorten time to market, reduce cost and improve quality, they are also subject to a series of pressures, of varying intensities, which in combination are unique to the specific set of circumstances in which the company finds itself. It is these pressures which act to modify any concurrent engineering implementation and ensure that each company must tailor concurrent engineering to provide its own specific solution.

In order to support the activity of tailoring concurrent engineering, this book develops a framework within which all of the influencing pressures acting on a company can be considered. The framework shows how these pressures can be categorized and then related to the elements which come together to form the concurrent engineering solution.

This book provides a series of case studies to illustrate the various pressures at work and the corresponding concurrent engineering solutions that companies adopt. In each chapter companies which have employed similar approaches to concurrent engineering are compared. Whilst all of the companies who have contributed case studies have considered many factors during the implementation of concurrent engineering, the reader will be directed to the particular areas where the majority of activity was focused. By these means the multitude of factors that determine a concurrent engineering solution will begin to be placed within a framework that helps to clarify the decision-making process.

C. J. Backhouse
N. J. Brookes

Acknowledgements

We would like to thank all of the contributors to this book for taking the time and effort to recount their experiences of implementing concurrent engineering. Without exception, the authors of the case studies found themselves in the typical position of today's manufacturing community – that of having too much to do and too little time in which to do it.

In addition, we would like to thank the Engineering and Physical Sciences Research Council, particularly the Control Design and Production Programme, which provided research funding of which this book is one result.

CJB
NJB

List of Contributors

Chris Backhouse is the Head of the Department of Manufacturing Engineering at Loughborough University of Technology. He has spent all of his professional career in the areas of product design and manufacture, initially in the field of high speed machine design and then later in design management. He was previously employed for a substantial period by Unilever Research, being involved in transferring best manufacturing practice between Unilever operating companies on a global basis. He left Unilever to join Loughborough University of Technology in 1990 in a joint role position whereby half his time was spent as Design Manager in Morris Mechanical Handling Limited, a Loughborough based manufacturer of cranes and lifting equipment, and the other half as a Senior Lecturer. In 1994 he moved to lecture full time at Loughborough University and in 1995 he assumed his current position.

Naomi Brookes is a researcher employed by the Department of Manufacturing Engineering at Loughborough University of Technology. She completed her Doctorate at Imperial College whilst employed on a Rolls-Royce Teaching Company Scheme where she had responsibility for the design and implementation of a computer-integrated inspection system for turbine blades. She was subsequently employed by Rolls-Royce in a variety of roles including shop floor and design office management, which led to responsibility for implementing a corporate approach to concurrent engineering with Rolls-Royce suppliers. She moved to Loughborough University of Technology in 1993 to progress her interest in the management of the product introduction process.

Ian Barclay worked for 20 years in industry with the Dunlop Company Ltd and was the Technical Manager of Dunlop GRG

Manchester from 1972 to 1978. In 1978 he joined the Polytechnic, Huddersfield, and amongst other activities set up the Technical Management Unit. He joined the Department of Industrial Studies at the University of Liverpool in 1984 to specialize in research on new product development and concurrent engineering. In 1991 he received a Leverhulme Trust Research Fellowship to study the 'Management of the new product development process' which included work with companies in the USA and Japan. He currently holds the BICC Cables Chair in Technology Management in the School of Engineering and Technology Management at Liverpool John Moores University.

Wendy Bowden is an Organisational Development Consultant within Measurement Technology Ltd based in Luton, Bedfordshire. Her experience ranges from production management of a large electronics company, with particular emphasis on the development of self-managing teams, to playing a key role in the introduction of concurrent engineering in the commercial aircraft industry.

Jane Burns is Operation Manager of Temco Ltd, Nottingham site, which is part of BICC Cables' Rod and Wire Operations. A Chartered Mechanical Engineer, her career to date with BICC Cables has included Manufacturing Engineer and Product Development Manager at Temco's Cinderford site. Her contribution to this book was written as a result of her work in this position, and contributed towards her MSc in Manufacturing and Technology Management.

Professor N.D. Burns is the Davy Professor of Manufacturing Systems at Loughborough University of Technology. His role is unusual in that he is also a Senior Manager at Morris Mechanical Handling Ltd, a leading crane manufacturer based in Loughborough. At the university he has research projects concerned with the behavioural aspects of performance measurement and the design of networked manufacturing systems. At the company he has been responsible for business change and improvement, primarily concerning the organization and management of the cellular-based systems within the business.

Dr Fiona Lettice has an MSc in Computer Integrated Manufacturing and a PhD in Concurrent Engineering Implementation. She has several years' experience in the energy and consultancy industries and is currently a Lecturer at Cranfield University, where her research interests are in change management with particular emphasis on teamworking and the product development process. She is now working on a research project which involves a best practice survey of concurrent engineering implementation in manufacturing companies in Europe and the United States and the development of a methodology for companies who wish to implement concurrent engineering. Fiona helped to set up and co-hosts a forum for a group of experienced concurrent engineering practitioners to regularly meet and discuss concurrent engineering-related issues.

Phillip Lewis is Assistant Dean at the School of Engineering, University of Derby, where he researches and teaches courses in Innovation and Design Management. He also leads the innovative BSc (Hons) Product Design, Innovation and Marketing course. Prior to entering higher education, he was involved for a number of years in international manufacturing management and consultancy at Lucas Industries plc. Much of this time was spent working on and managing projects concerned with improving the new product introduction process. This was achieved through the implementation of best practice design management principles such as concurrent engineering.

Stuart Marshall is Development Manager for Storage Subsystems at IBM, Havant, and is responsible for developing high-performance disk drive subsystems for the open-systems market and advanced technology to support future growth opportunities. He took this position in 1992 following a technical marketing role supporting storage marketing. Formerly he has worked on optical and magnetic disk development. He joined IBM in 1974 as a chip designer and has held various engineering and management positions in disk drive development. He was appointed a visiting fellow at Bournemouth University following several years of joint work in the field of concurrent engineering. He has recently been chairman of the DTI sponsored Time to Market Association.

John McAllister graduated from Durham University in 1961 with an Honours degree in Mechanical Engineering. Throughout his career he has been involved with servohydraulic systems in applications ranging from aircraft flying controls to industrial robots. Since 1972 he has been with Instron Ltd working on the development and application of servohydraulic testing machines and structural testing systems. He has held several positions within Instron and is currently Control Group Manager. John has a long-standing interest in improving the product development process and in 1993–94 chaired an inter-departmental Task Force within Instron developing ways in which greater use could be made of concurrent engineering methods. He is also the Instron representative on Loughborough University's SIMPLOFI project, of which Instron is an industrial partner.

Mike Meadowcroft joined ICL in 1974 and has held engineering and managerial positions in manufacturing and computer development divisions. Over the last seven years he held strategic responsibility for the CAD Tools and Engineering Databases used by the Client-Server Systems Division of ICL. In 1991 he received an MSc in Information Systems Design and Management from Kingston University, and is a Member of the British Computer Society. Since 1994 he has managed an engineering department within Design to Distribution Ltd (an ICL company) providing product development services to design teams in many organizations.

Professor Sa'ad Medhat is Head of the newly created Office of Research at Bournemouth University with a campus-wide responsibility for supporting, coordinating and promoting research and consultancy across all disciplines. He worked as a Computer Analyst at Philips, and subsequently was Technical Director of Data-Tech International before joining Bournemouth University. Sa'ad continues with active programmes of research and research supervision in the areas of concurrent engineering, managing processes and technology and innovation. He has edited 13 journals and books and has published over 40 papers nationally and internationally.

Bruce Norridge was born in Johannesburg, South Africa and spent the first 14 years of his career as a structural engineer working in pure design. At the age of 38 he went back to University to take the South African Production Engineering Institute Exams. At the age of 42 he graduated *summa cum laude* and immediately went into production management. He discovered his forte was fixing ailing companies, hence his interest in people and the mechanism of management techniques used in the process of continuous change. Bruce moved to the UK in 1992 to effect changes in the business he is presently involved in. He is a registered Professional Technologist in Engineering, S.A., a Registered Professional Production Manager, S.A. and is a Fellow of the Production Management Institute of S.A.

Tim Pegg spent a period on an exchange program at the Massachusetts Institute of Technology (MIT) in Cambridge, USA, after which he graduated from the University of Manchester Institute of Science and Technology (UMIST) with a BSc in Electrical and Electronic Engineering. In 1983 he joined Marconi Instruments Ltd working on graphical display technology. The next two years were spent working on digital hardware and software design for the 6300 series Programmable Sweep Generator. This was followed by a period in Project Management, during which time he led the team in the development of the 6200 series Microwave Test Set. After a brief period as Production Engineering Manager he was appointed Product Group Manager for the Microwave Group at Marconi Instruments Ltd in 1992.

Jenny Poolton was born in Birmingham, England in 1957. She received the B.Sc (Hons) Psychology degree from Warwick University in 1990, and the PhD degree from the University of Liverpool in December 1994. The thesis is entitled 'Concurrent Engineering Establishment: A Framework Proposal'. Latterly, Jenny is employed as a Research Fellow in the Department of Industrial Studies at Liverpool University. Her main research interests concern how firms improve their new product development performance. She is a member of the British Psychological Society.

Graeme R. Pratt received an Honours degree in Mathematics from Brunel University in 1963 and then worked in the nuclear, steel and defence industries before joining the Faculty of Computing Sciences at the University of Alberta, Canada. He joined IBM in 1968 as a systems engineer before moving into new product development of mainframe, communications and graphics products. He then held senior management positions in product development and international marketing in Paris, New York and the UK. After retiring from IBM in 1993, Graeme became Development and Industrial Liaison Manager in the Office of Research at Bournemouth University where he also lectures in international marketing, business management and concurrent engineering.

Jim Rook is a researcher with the Department of Electronics at Bournemouth University (working in collaboration with IBM, Havant). He is currently undertaking a PhD in the concurrent engineering field, his thesis title being 'Defect Analysis and Predictive Modelling for Efficient Electronic Production Development in a Concurrent Engineering Environment'. Jim received a BEng (Hons) in Manufacturing Engineering from Brunel University and an MSc in Concurrent Engineering from Bournemouth University.

Oliver Towers graduated from Cambridge University in 1977, after which he carried out research at the Welding Institute. He joined Rolls-Royce in 1984 as a Research Engineer, and in 1986 he formed a team to develop computer models of manufacturing processes. From 1990 Oliver was employed as Development Engineering Business Manager before leading a team which re-designed the Aerospace Group's approach to product development with the purpose of reducing unit costs. He was responsible for implementing the team's proposals as Head of Process Improvement (Procurement). In 1995 he was appointed EU Relations Executive, based in Brussels.

1
Concurrent Engineering: Where it has Come from and Where it is Now

Naomi Brookes and Chris Backhouse

Introduction

The autumn of 1988 saw the Rover Group launch their new four-wheel drive vehicle, the Discovery. Pitched to compete with the Shoguns, G-Wagens, Land Cruisers, Patrols and Troopers built in Japan and Germany, it gave consumers a new choice in this rapidly developing and highly competitive vehicle category. Just one year later sales of the Discovery had outstripped those of all its rivals to become the best-selling model in the four-wheel drive UK market sector. It has now gone on to become one of the most successful of all Rover models. What was remarkable about the launch of the Discovery was that it was introduced to the market in half the best time to market previously achieved by Rover. Not only that, it also came in on budget and a week ahead of schedule.

The Discovery's continuing success in the marketplace has demonstrated that its appeal to the customer was not simply based on its novelty but on a high quality product design derived from a clear view of customer requirements. The design of the vehicle was not seen as being compromised by the accelerated development process – far from it. The high quality of the final design was in fact a direct consequence of the rapid product development which required that all aspects of the vehicle were undergoing constant evaluation.

Tony Gilroy, Land Rover's chief executive at the time, put the Discovery's success down to the use of concurrent engineering in improving the performance of the company's product introduction process. This new approach to organizing the way new vehicles were developed focused around the concept of improving communications between all personnel having a role in the new design. The Discovery was introduced using a multidisciplinary team which strove to remove the organizational boundaries that had fettered development activities in the past.

The team members were deliberately co-located as a fundamental of the new approach. This was in direct contrast to previous practice which involved separately locating each company function within definable boundaries. The concurrent engineering team was entirely responsible for setting up its own *modus operandi* once its targets had been specified. It performed activities in parallel wherever possible, a task much eased by the close proximity of all team members, and employed the full range of quality techniques and computerized tools available at that time.

Chrysler's experience

Rover's success in using a concurrent engineering approach, or simultaneous engineering as it is sometimes known, reflects the experience of many practitioners. Chrysler used the technique to very great effect in recovering from record quarter-year losses of $664 million reported in February 1990. The company had slipped to fifth position in terms of US market share due to an outdated and uninspiring product portfolio. It owed any sales it did have to a core of highly loyal aging customers who in the words of Lee Iacocca, the charismatic chairman of the time, were 'between 55 and dead'. In the Management Brief 'Crunch at Chrysler' published in *The Economist* in November 1994, another quote describing the situation in which the company found itself was 'the only question at that time was which would die first, these loyal customers or the company'.

The contrast between Chrysler of 1990 and its situation just four years later could not have been more dramatic. At a London conference 'Meeting the Challenge of Global Manufacturing – World Scale Collaboration' in October 1994, sponsored by the Department of Trade and Industry, a representative from Toyota spoke of being 'out-engineered' by Chrysler's new mid-range saloon car – the Neon. He felt that the lower component count of the Neon gave it a distinct competitive advantage over its rivals through its being an inherently lower cost car to produce.

A report issued after Toyota had performed a detailed strip-down of the Neon said that it incorporated 'designed-in savings unprecedented in an American car'. The company which, four years earlier, had been regarded as incapable of producing a car suited to the modern consumer had managed to outstrip all its competitors, in both America and Japan, to develop the best car in its class. The success of the Neon and its sister range of small to medium-sized cars was put down to one vital ingredient – concurrent engineering.

Chrysler brought together people from all of the functions involved in product development into what it termed 'platform teams'. These teams cooperated in every stage of the product's development from conception through to delivery. No longer were personnel compartmentalized into groups with the design progressing sequentially between the various functions. Instead, the platform teams were multifunctional, containing all of the various engineering functions together with marketing, sales and purchasing. Personnel within the teams worked concurrently by ensuring that their particular knowledge was taken into account by other members of the team at the first opportunity. Manufacturing engineers were involved from the earliest stages to ensure that each individual component could actually be manufactured and assembled into the final product at an acceptable cost. Purchasing staff were able to influence the design from day one, advising on the availability and cost of materials and parts, and offering alternatives where trade-offs between cost and function were being debated.

All of this was in direct contrast to the development of previous vehicles where designs had been progressed in a sequential manner based on the philosophy of 'we design it – then you make it'. Under the new arrangement, illustrated in Figure 1.1, all downstream activities were involved from the earliest possible stages in the process.

The new product range was developed in three years with a third of the engineering resource that would traditionally have been used. From a loss of $538 million in 1991 Chrysler recovered to make net profits of $24 billion in 1993. The role of concurrent engineering in developing the product range to fuel this recovery was central to the company's success.

Concurrent engineering – a short history

As with every good idea, concurrent engineering has been around in one guise or another for a very long time. The need for accelerated product development and its achievement through parallel tasks are not new. Henry Ford was instrumental in introducing to design teams the concepts of designing for manufacture and assembly during the development of the Model T automobile. Described in many books, including Burlingame's short biography *Henry Ford*, in 1907 Ford expressed his vision of the future in the following terms:

> I will build a car for the great multitude . . . of the best materials by the best men . . . after the simplest designs that modern engineering can devise.

This view of developing automobiles which suited his customers' requirements at a price they could afford was an all pervading philosophy throughout Ford's long and distinguished career. Although competitive pressures from fellow automobile manufacturers sometimes caught him unprepared, in the following decades it can be justifiably stated that he was the primary driving force in developing the American mass production approach to consumer goods manufacturing. He was instrumental in progressing highly successful projects throughout the 1920s and 1930s, continuously looking for ways of reducing the time to market whilst keeping prices down and quality up.

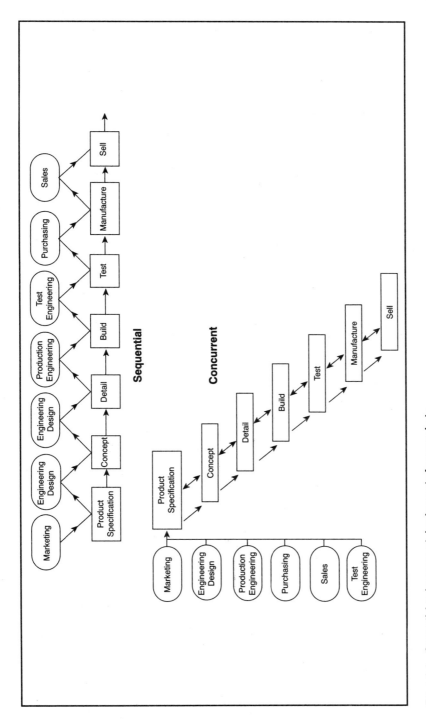

Figure 1.1 **Sequential and concurrent development of new products**

War years

By the time America entered the Second World War the approach to designing new products using multifunctional teams was considered the norm since competition based on price and new features was so fierce. Ford, and other companies such as Chrysler, provided the ideal platform from which the American armaments industry could increase production after December 1941. Their ability to develop new products which could be quickly and easily manufactured was already well established for the civilian market. It was relatively straightforward to apply the same basic principles to the development and production of tanks, aeroplanes and ships. The result was that, although America entered the war lacking a dedicated armaments industry, by 1945 it had produced more weapons of war than all other participants combined.

The situation in the UK at that time was somewhat different. The armaments industry was already well established and, whilst this brought some immediate benefits, it did not facilitate such a rapid rise in output as that experienced in America. Accepted practices had to be broken down before the full benefits of teamwork could be realized. Although the output of aircraft, for instance, rose significantly during the war years, it was hindered by pre-war designs which had focused almost exclusively on performance to the detriment of manufacturability. The Spitfire, for example, outflew the equivalent German Messerschmidt but required three times as many man-hours to produce.

Whilst the UK had to live with the consequences of its pre-war designs, the development of new designs encompassed a changed set of priorities. A focus on design for manufacture and teamwork resulted in dramatic improvements in the UK industry's ability to produce armaments. Rolls-Royce in 1943 employed precisely these techniques to deliver the first Derwent gas-turbine aircraft engine in five months from initial specification to successful prototype. The company repeated this success in 1944 with the introduction of the significantly more sophisticated and powerful Nene gas-turbine engine following a six-month product introduction process. Similar achievements were being recorded in new aircraft, tank and other armoured vehicle designs throughout the UK armaments industry.

As part of this rapid product introduction the need for close collaboration between disciplines was also recognized. Both the airframe manufacturers and component suppliers were closely involved in the development programme on a daily basis to ensure coordination of activities. Design engineers and manufacturing engineers, who had previously worked sequentially within independent units, cooperated closely to ensure that the design not only achieved its performance specification but could also be manufactured quickly.

The post-war boom

In the early post-war years when industry was reverting to supplying the civilian consumer market, many industries retained the same focus on product development as before. The principles of good product development were often remarkably similar to those used to describe concurrent engineering today. Kodak, for instance, explicitly stated the key points for successful product design and development within its organization as:

1. Close interaction between commercial needs and the work in the laboratory is essential.
2. Product innovation and process innovation are intimately related.
3. Progress occurs through organized effort and teamwork; collaboration across functional boundaries is imperative.

However, as the industrialized world experienced the growth decades following the Second World War, the major manufacturing industries evolved into giant bureaucracies. As demand for modern consumer products began to outstrip supply, companies found that they could remain highly profitable without having to work too hard. Under these circumstances it is not surprising that they relaxed and changed their emphasis. The lessons for rapid and efficient product development were lost as companies divided themselves into specialist functions and began to follow sequential design processes.

This move to sequential operations where individual designers contributed their effort and then passed the design 'over the wall' for the next set of inputs was ideally suited to these new organizations. All such bureaucracies were more intent on monitoring their internal processes than with their external customers. All activities were easily defined and compartmentalized and the fact that this led to inefficiency was irrelevant. The marketplace had the capacity to soak up highly profitable products as fast as they could be produced. Under these conditions there was no incentive to move towards 'leaner' organizations nor to develop more efficient operating procedures and business processes.

The rise of quality

Although the major corporations were becoming less efficient in the post-war years, new concepts on which future competitiveness would depend were still being developed. These were not necessarily implemented on a wide scale, but key ideas regarding how to develop new products were emerging. Unfortunately for the developed world the most responsive audiences for the new ideas were in Japan, a country which was thought at that time to have little potential in terms of world manufacturing competitiveness.

American quality gurus such as W. Edwards Deming and Joseph M. Juran rose to prominence in the reconstruction of Japan's manufacturing industry during the late 1940s and early 1950s. Their approach to quality in the early years consisted of the application of statistical techniques aimed at minimizing production variability. However, they extended their activity to encompass systematic methodologies for problem solving and introduced many of the concepts which came later to be identified with the concept of Total Quality.

Juran developed a 'Quality Planning Road Map' which began with the identification of customers and ended with a proven process capable of producing the designed product. Deming was convinced that senior management involvement was central to the success of quality improvement programmes, and he produced a series of 14 fundamental points to explain how to succeed. Whilst these range over a number of areas, from strategic to tactical, it is interesting to note the close relationship they hold to concurrent engineering. For example, one of Deming's 14 points described in his book *Out of the Crisis* states:

> Break down barriers between departments. People in research, design, sales and production must work as a team to foresee problems in production and in usage that may be encountered with a product or service.

Whilst some of the 14 points are philosophical issues such as the one quoted above, others are methods for improvement, such as instituting a vigorous education programme. However, all of the points he raised form important elements of the overall concurrent engineering philosophy. The concept of supplying the customer with the right product, at the right time and at an affordable price was central to Deming's philosophy and is similarly central to concurrent engineering.

Competition from Japan

The steady and substantial growth in the market's capacity for new products during the 1950s and 1960s protected the major industries in America and Europe from the consequences of their inefficiencies. Japan, a country which had never ceased to believe in efficient manufacturing, emerged as a major new competitor. It had listened closely to many of the pre- and early post-war American theories on effective design and manufacture and had refined them to new levels. Initiatives in the way labour was organized, the overriding drive for quality, the focus on manufacturability all stemmed from ideas developed many years earlier in the American mass production system. However, the concepts now seemed new and revolutionary to western manufacturing industries who had become used to feeling unassailable in their market dominance.

These new concepts for efficient manufacturing were at first thought by western manufacturers to be individual and unrelated initiatives. Specific targets, such as improving product quality, reducing overall costs, designing products that the customer really wanted and reducing inventory costs, were seen as isolated initiatives. Those western companies with foresight monitored their Japanese rivals and as new techniques emerged they were quick to put them into practice in their own companies. However, as they only concentrated on the individual techniques they failed to appreciate the bigger picture – that the new competitive weapon of the late twentieth century was to be rapid new product introduction at low cost and with inherent high quality.

Multidisciplinary teams

Teams were being specifically identified as one of the approaches to rapid product development as early as 1982. Peters and Waterman, in their book *In Search of Excellence,* made famous the 'skunkwork' teams operating within Minnesota Mining and Machinery Co. (3M). These multidisciplinary teams were responsible for new product introduction. Their efforts met with considerable success and 3M claimed that 25 per cent of its revenue came from products less than five years old.

Peters and Waterman were not alone in their perception of the importance of teams in product development. Many writers on organizational strategy, such as Gerald Mathot and James Quinn, were echoing these ideas at the same time. The use of teams for product development received another boost in 1986 with the publication of the widely acclaimed *The Superteam Solution* (Hastings et al, 1986). The favourable disposition towards team-based organizational structures that existed at the time fed directly into the establishment of the concurrent engineering environment.

Transforming change

The 1980s were characterized by a growing realization of the importance of new product introduction to a company's success. This was prompted by the rapidly increasing rate of change that businesses were beginning to experience. Rosabeth Moss Kanter, in her book *The Change Masters* first published in 1983, described this change in the following terms:

Business organisations are facing a change more extensive, more far reaching in its implications and more fundamental in its transforming quality than anything since the modern industrial system took place. These changes in the business environment come from several sources: the labour force, patterns of world trade, technology and political sensibilities. Each of these has by itself changed significantly at other times. The present situation is unusual not only in that each is undergoing transforming changes, but that the changes are profound and that they are occurring together.

The changing nature of both technology and markets led to the situation where individual product life-cycles were rapidly reducing. To remain competitive companies needed to introduce more and more new products. A survey by consultants Booz Allen in 1982 had already highlighted the contribution that new products were going to make to company revenues. Yet again, this appears to be a trend that was picked up and quickly acted on by the Japanese but less so by most western manufacturers.

In Japan, one of the prime principles of management was that new products must be constantly introduced to continue the cycle of investment, cost reduction, price reduction and market share increase. A 1984 survey of Japanese chief executives gave their primary role as increasing market share and their next goal as increasing the ratio of new to old products. However, when asked how they would achieve their first objective, they replied that they would seek to introduce an increasing proportion of new products. In the mid-1980s, the strategic importance of new product introduction began to be more widely promulgated in the west. A series of articles began to appear in the *Harvard Business Review* – the prestigious journal published by the Harvard Business School which is often at the forefront of business trends. For example, in 1988 the *Harvard Business Review* printed an article by Daniel Whitney, a well-known design researcher and consultant, who said that:

> Design [of product and process] is a strategic activity whether by intention or default. It influences flexibility of sales strategy, speed of field repairs and efficiency of manufacture. It may well be responsible for the company's future viability.

This viewpoint was echoed the same year by Professors Hayes, Wheelwright and Clark of the Harvard Business School:

> One can build a competitive advantage through superior manufacturing, but sustaining it over time requires comparable skills in creating a new stream of products and processes.

The importance of new product introduction is now widely accepted. In its 1989 report on major issues facing businesses in the 1990s, PA Consulting Group identified this activity as one of its key responses in facing the challenges of the decade. The UK government also strongly identified this theme in its 'Managing for the 1990s' initiative.

Benchmarking product introduction performance

In 1985, the Massachusetts Institute of Technology (MIT) had begun a five-year, five million dollar international study into the future of the automobile industry. This was summarized in the book *The Machine that Changed the World* (Womack *et al.*, 1990). The project looked at all aspects of the industry but devoted a significant amount of effort to comparing product introduction in Japan and in the West. The researchers considered the work of, amongst others, Kim Clark and Takahiro Fujimoto of the Harvard Business School, who in 1991 published their work under the title *Product Development Performance*. Clark and Fujimoto had performed a parallel study that focused particularly on product introduction. A number of key lessons were gained from these studies:

- the importance of leadership to a project
- the importance of teamwork
- the importance of communication
- the importance of simultaneous (concurrent) development.

The MIT study was probably one of the most comprehensive comparisons of companies operating within a single industry sector and it also highlighted many of the issues concerning all industry sectors. It particularly focused on approaches to managing the process of developing new products and evaluating product introduction effectiveness.

The global market

Until recently most companies were shielded from pressures to change by geography, protected technologies and loyal customers. However, the new global market has transformed the environment in which manufacturing industry exists. Prior to the early 1990s only a few manufacturing industry sectors could consider themselves as existing in a global market. Rapid changes in the market itself have created financial imperatives forcing new product development.

The characteristics of the new global marketplace are now recognized. Clark and Fujimoto identify the global market as containing three fundamentally different characteristics from that which preceded it:

- International competition is becoming more intense as large numbers of sophisticated industries with highly competitive cost structures are being established in developing countries.
- Sophisticated consumers with high levels of expectations are emerging in fragmented markets around the globe. This is stimulated by the rapid development of communications capabilities, with particular relevance to those communication media which are primarily financed through advertising.
- The readily achievable transfer of technology is resulting in companies experiencing a reduced 'window' in which they can reap the benefits of any technological advance. Competitors are now capable of almost eliminating learning curves in many industry sectors.

The consequences of these three pressures are that companies are competing in an environment where product life-cycles are reducing, customers are demanding lower-cost products and quality is taken as a prerequisite. The cumulative effect on any company is to reduce profitability through shortening life-cycles, and the only realistic remedy is the more rapid development of new products.

The emergence of concurrent engineering

The term 'concurrent engineering' emerged as a description to encompass the whole philosophy of providing competitive advantage through rapid new product introduction following a study by the US Defense Advanced Research Projects Agency in 1987. It was intended to demonstrate through its very title that the processes necessary to develop new products must be conducted concurrently. Only by this approach could companies achieve the targets associated with cost, quality and time that the new markets were demanding. Within the umbrella term of concurrent engineering the isolated improvement initiatives on which companies had previously focused were identified as a series of building blocks which could be brought together.

Thus by 1987 the realization that new product introduction would be a key industrial criterion of the future had spawned a term to describe the processes required to achieve competitive performance. Many larger companies could now look back into their recent history and recognize that the reorganizations they had initiated to counter the Japanese threat had in fact been based on the fundamentals of concurrent engineering. The financial threat that they had perceived through loss in sales had forced them to look very closely at how the competition from the Far East had developed high quality products so quickly. They had recognized that their very survival depended on quickly restoring their financial well-being. They had come to realize that a policy of no improvement would result in rapid, and probably terminal, profit decline.

Concurrent engineering defined

Publications that referred to the concept of concurrent engineering began to appear in earnest in 1987. These focused on the use of simultaneous engineering, as it was then called, in the American automobile industry. Early successes included the development of the Ford Taurus model which achieved a dramatic reduction in lead-time in comparison to previous projects.

In 1987, the US Defense Advanced Research Projects Agency (DARPA) set up a working group of academics, industrialists and government specialists to examine the implications of simultaneous engineering for defence sourcing. The working party wholeheartedly supported the concept but rechristened it 'concurrent engineering' better to reflect the concept of parallel processes. They also provided the manufacturing community with a definition:

> Concurrent Engineering is a systematic approach to the integrated concurrent design of products and their related processes including manufacturing and support. This approach is intended to cause the developers from the outset to consider all elements of the product life-cycle from conception through disposal including quality, cost, schedule and other user requirements.

The principal recommendations of the working group were that the Department of Defense should take positive steps to encourage the use of concurrent engineering in weapons system acquisitions and, in parallel, that it should take steps to support the cultural, managerial and technical changes necessary to obtain the full potential benefits of concurrent engineering. This led to the setting up of the DARPA Initiative into Concurrent Engineering (DICE) technology development programme.

Extending the definition

By the end of 1991 about $60 million had been invested by the US government under the auspices of the DARPA initiative. The introduction of concurrent engineering to the defence, automobile and aerospace industries acted as a trigger at the start of the industrial supply chain. Its use spread rapidly through suppliers and subcontractors. However, whilst DARPA produced a usable definition of concurrent engineering it did not necessarily satisfy all of the requirements for implementation. At one level it failed to explain the *need* for concurrent engineering. Even though we now have the benefit of hindsight it is fair to say that during the period that DARPA was deliberating most of manufacturing industry was blissfully unaware of the level of importance new product introduction was going to attain in the 1990s. In particular, whilst those companies at the pinnacle of the supply pyramid such as Boeing and Ford recognized the importance of concurrent engineering, the smaller component manufacturers at the base of the pyramid were less well informed.

In this situation it was necessary for the uninitiated to be provided with a definition of why concurrent engineering was indispensable, rather than of how it operated. Robert Creese and Ted Moore, in a 1990 article discussing the strategic benefits of concurrent engineering, defined it as:

> a management philosophy dedicated to the improvement of customer satisfaction through improved quality, reduced costs and faster product development.

At the other extreme practitioners were seeking help in implementing the detail of concurrent engineering. Whilst they could now understand why concurrent engineering was necessary and also what the targets were, they had little knowledge of how to create their own concurrent engineering environment. One of the most widely used guides to implementation was produced by the writers John Hartley and John Mortimer, who defined how concurrent engineering can be implemented via the following elements:

- multidisciplinary task-forces
- the product defined in customer terms and then translated into engineering requirements
- design of process parameters
- design for manufacture and assembly
- concurrent development of product, manufacturing process, quality control and marketing

A situation was therefore developing where definitions for concurrent engineering were being created in terms of why, what and how – why it is needed, what it is, how it is implemented. It may be true that during this period these distinctions were not so clear to all those looking to implement concurrent engineering. However, major companies who did have a clear view of the future were instrumental in driving through change. First-, second- and third-tier suppliers were forced to adopt this new way of developing products, even if at the beginning the purpose of the approach was not completely understood by smaller companies providing individual components.

The aircraft industry

Whilst the DARPA initiative was primarily targeted at defence procurement, its effects rapidly migrated to related industrial sectors. As has previously been discussed, the automobile industry quickly adopted the concurrent engineering approach. However, the aircraft industry was just as ready to adopt the new ways of working. Boeing's latest and largest aircraft, the Boeing 777 wide-bodied jet, made significant use of concurrent engineering techniques and is now enjoying, as a result, major benefits in terms of reduced design changes, reduced time to market and reduced costs.

Boeing created 238 teams to work on the 777 development programme, formally including customers and suppliers as members of the development teams. Extensive use was made of modern computer aided engineering technology to enable both design and analysis to be carried out within an environment which actively encouraged contributions from all team members at the earliest possible stages.

Boeing suppliers were not left behind in the adoption of this new approach to product development. At the simplest level Boeing's targets for reduced lead-times required that component manufacturers should be equally capable of developing their new products on drastically reduced timescales. However, it also implied that the development of component parts would have to be much more closely integrated into the overall design process. This was necessary to ensure on-time delivery of quality products which would function on a right-first-time basis.

It was no longer acceptable to deliver components requiring rework or modification. The development programme of the new aeroplane required that, on delivery, any component would have to comply entirely with the agreed specification. This in turn demanded considerably more effort than had previously been given to specification development. Only by rigorously defining all the design requirements of each component could Boeing ensure that all the multitude of activities involved in development were working towards the common goal. It was through this attention to detail that the introduction of concurrent engineering was ultimately successful in achieving the targets Boeing had set itself.

Concurrent engineering moves down the supply chain

Stimulated by the activities of Boeing the major suppliers followed rapidly to introduce this new approach to product development. Concurrent engineering was used by both the two major engine manufacturers, General Electric and Rolls-Royce, in order to achieve dramatic reductions in development lead-times. By increasing right-first-time engineering through intensified design collaboration with Boeing they ensured that the aircraft manufacturer was able to maintain its schedule for the introduction of its new product.

At the next level down the supply chain a similar process was taking shape. The component suppliers to both Rolls-Royce and General Electric experienced pressures to develop their products more quickly with minimum redesign and in closer collaboration with their customer. Just as Boeing had demanded the achievement of new targets from its engine manufacturers, so the next level of suppliers were similarly pressurized. And so the process of introducing concurrent engineering continued on down the supply chain until every company involved in the project had been affected.

Wherever they were in the supply chain each company was required to set shorter timescales for their activities and to collaborate on a level rarely seen in the past. Personnel from component companies were attached to the major subassembly companies in order to ensure that specifications, and ultimately component designs, were developed with a full knowledge of cost, quality, function and time.

Component suppliers were not slow to see the competitive potential of this way of working in other product lines. Although the concurrent engineering stimulus may not initially have affected all of their operations, the new working practices were rapidly transferred throughout complete organizations. This in turn stimulated new supply chains rapidly to reconfigure their relationships and ways of working. As a consequence, vast swathes of manufacturing industry quickly became aware of, and began implementing, concurrent engineering.

Concurrent engineering across sector boundaries

This development of awareness and implementation of concurrent engineering within certain sectors of manufacturing industry was repeated elsewhere, and in certain cases occurred prior to the examples already cited. Industry sectors associated with electronic consumer goods, such as video recorders and personal computers, were well aware of the pressures to accelerate the introduction of new products and reacted accordingly. Major companies such as Philips, Hewlett-Packard and Motorola have had to respond to rising competitive pressures from companies based in the Pacific Rim and have introduced approaches to developing new products based on the principles of concurrent engineering. As a consequence, the component suppliers have had to respond in a similar manner to that described for Boeing's suppliers. Closer collaboration with major original equipment manufacturers and a more rapid turnround in development projects is now the 'order of the day' for all companies contributing to these supply chains.

Educational needs

It is now rare to find any sector within manufacturing industry which is not experiencing pressure to introduce new products at an increasing rate whilst simultaneously aiming for moving targets in terms of cost and quality. Whether the products being supplied are relatively mundane mass produced parts or highly complex one-off capital plant items, the pressure to accelerate product introduction is ever present. Major multinational companies are being driven by the growing pressures of an increasing number of competitive products and this pressure is being transmitted throughout manufacturing industry.

As has already been illustrated, smaller companies operating lower down supply chains have been influenced by major original equipment manufacturers to implement concurrent engineering. However, this drive to implement has not always been matched by the knowledge of how to achieve best results. Concepts which worked well in large multinationals did not always transfer so easily to the smaller component supplier. A multitude of 'platform teams' as created by Chrysler is not so easily repeated in a company of 50 or 60 employees. Indeed, surveys of manufacturing industry during the early 1990s discovered that whilst many companies were introducing concurrent engineering, their understanding of the concept was limited. They lacked knowledge in developing the form of concurrent engineering most appropriate to their circumstances. They did not always appreciate that successful implementation in one company could not necessarily be used as a prescriptive template for a second company.

The implication is that significant clarity of thought is required in order to identify the correct form of concurrent engineering appropriate for a specific set of circumstances. It is this concept that the following chapter and subsequent case studies aim to address.

Bibliography

'Issues of Co-Operative Working in Concurrent Engineering', *IEE Digest*, 1994/177, 1994.

'The 1993 Manufacturing Attitudes Survey: a Management Summary', London: Computervision, 1993.

Booz-Allen & Hamilton Inc (1982) *New products management for the 1980s*, New York: Booz-Allen & Hamilton.

Burlingame, R. (1957) *Henry Ford*, London: Hutchinson.

Carter, D.E. and Barker, B.S. (1992) *Concurrent Engineering: The Product Development Environment for the 1990s*, Reading, Mass.: Addison Wesley.

Clark, K. and Fujimoto, T. (1991) *Product Development Performance:*

Strategy, Organisation and Management in the World Auto Industry, Boston, Mass.: Harvard Business School Press.

Creese, R.C. and Moore, T.L. (1990) 'Cost Modelling for Concurrent Engineering', *Cost Engineering*, **32**, (6), 23–26.

Deming, W.E. (1982) *Out of the Crisis: Quality, Productivity and Competitive Position*, Cambridge: Cambridge University Press.

Hartley, J. and Mortimer, J. (eds) (1990) *Simultaneous Engineering*, Dunstable, Beds: Industrial Newsletters.

Hastings, C., Brixby, P. and Chaudry-Lawton, R. (1986) *The Superteam Solution*, Aldershot: Gower.

Hayes, R., Wheelwright, S. and Clark, K. (1988) *Dynamic Manufacturing: Creating the Learning Organisation*, New York: Free Press.

Juran, J.M. (1989) *Juran on Leadership for Quality*, New York: Free Press.

Kanter, R.M. (1992) *The Change Masters: Corporate Entrepreneurs at Work*, London: Free Press.

Kusiak, A. (ed.) (1993) *Concurrent Engineering: Automation, Tools and Techniques*, New York: Wiley.

Mathot, G. (1982) 'How to Get New Products to Market Quicker', *Long Range Planning*, **15**, (6), 20–30.

Nevins, J.L. and Whitney, D.E. (1989) *Concurrent Design of Products and Process*, New York: McGraw-Hill.

PA Consulting Group (1989) *Manufacturing into the late 1990s*, London: HMSO.

Parsaei, H.R. and Sullivan, W.G. (1993) *Concurrent Engineering: Contemporary Issues and Modern Design Tools*, London: Chapman & Hall.

Peters, T. and Waterman, R. (1983) *In Search of Excellence*, New York: Harper & Row.

Pye, A. and Mynott, C. (1993) *UK Product Development Survey*, London: The Design Council.

Quinn, J. (1985) 'Managing Innovation: Controlled Chaos', *Harvard Business Review*, May–June, 73–84.

Ranky, P.G. (1994) *Concurrent/Simultaneous Engineering: Methods, Tools and Case Studies*, Guildford: CIMWare Ltd.

Scherkenbach, W. (1986) *The Deming Route to Quality and Productivity: Road Maps and Road Blocks*, Washington: Mercury.

Shina, S. (1991) *Concurrent Engineering Design for Manufacture of Electronic Products*, New York: Van Nostrand Reinhold.

Shina S. (ed.) (1994) *Successful Implementation of Concurrent Engineering Products and Processes*, New York: Van Nostrand Reinhold.

Syan, C.S. and Menon, U. (1994) *Concurrent Engineering: Concepts, Implementation and Practice*, London: Chapman & Hall.

Turino, J. (1992) *Managing Concurrent Engineering: Buying Time to Market*, New York: Van Nostrand Reinhold.

Walton, M. (1986) *The Deming Management Method*, New York: Perigee.

Whitney, D. (1988) 'Manufacturing by Design', *Harvard Business Review*, July–August, 83–91.

Womack, J., Jones, D. and Roos, D. (1990) *The Machine that Changed the World*, New York: Rawson Associates.

2
Variety within the Product Introduction Process

Chris Backhouse and Naomi Brookes

The series of case studies described in this book will illustrate the fact that there are many different routes to implementing concurrent engineering. For the electronics manufacturing service supplier D2D described in Chapter 4, the concept of concurrent engineering consists primarily of an IT (information technology) infrastructure which comprises rapid communication and management of engineering data between geographically distributed sites. Virtual teams of cooperating engineers who rarely, if ever, physically meet can be established to progress highly complex product development assignments. In contrast, Morris Cranes, a manufacturer of capital plant described in Chapter 6, emphasizes the need for physical co-location of personnel to ensure that the vital element of communication is maintained at the necessary level.

Both of these two companies have common pressures on them to reduce product introduction time and costs whilst maintaining or improving quality. In many ways their approach is very similar, in that they are both aiming to improve communication between personnel involved with a project in order that design and manufacturing concepts can be taken into account as early as possible in the process. However, in their route to developing concurrent engineering the two companies clearly differ. The need for D2D to integrate its design activities on a global scale is not repeated within Morris Cranes who operate from a single design and manufacturing site.

Significantly, then, we can see that the response of two companies to the pressures of a competitive marketplace leads to very different processes involved with product introduction. Both of these companies consider that they are implementing concurrent engineering and they have long-term plans to develop the implementation even further. Their approaches to concurrent engineering do have several features in common, but crucially the two companies have different views on how to distribute priority between these features.

Such differences can be identified in all the case studies included in this book, reflecting the varying priorities for each company. The Rolls-Royce case study, for example, describes the emphasis being placed on supplier integration and the requirement for closer liaison between design engineers and the component producers. The Lucas Aerospace Actuation Division and Marconi Instruments case studies report on the reorganization of company structures to provide a more focused approach to product introduction. The Measurement Technology case study emphasizes the people aspects of concurrent engineering, whilst the Temco study focuses on core competences. For a company such as IBM Havant with a sophisticated IT structure the priority is to develop an integrated set of design tools, whilst Instron make the case for retaining elements of a functional organization to maintain core expertise.

The cyclical nature of implementation

It is evident that the priority each company gives to various aspects of a concurrent engineering implementation will vary in a cyclical way. This cyclical shift can often replicate cyclical market forces. The behaviour of many industrial sectors is recognized as following a repeating pattern. Perhaps the most well known is the chemicals industry, which experiences periods of high margins followed by pressures on costs as new capacity comes on-stream. The cycles that such companies experience relate directly to the time it takes to establish new production facilities. Similar, if somewhat less predictable patterns affect most companies and it is not surprising that they shift their priorities from, say, maximizing output to minimizing costs. Under these conditions the priorities that any concurrent engineering implementation must address will steadily change. Whilst global competition is requiring a more rapid response to changing priorities, it is a rare company that sees a consistent set of product objectives for a long period.

Secondly, there is an internal shift in priorities. Each of the companies in the case studies has described their priorities of the day. In addition, they have described the activity related to satisfying those priorities. Once the identified priority has received the necessary 'treatment' then it is inevitable that the returns on any further resource applied will begin to diminish. It is therefore to be expected, and indeed required, that a new set of priorities will be identified which will direct resource into a slightly different direction.

It is these two sets of influences – the external and the internal – that are creating a shifting cycle of priorities resulting in many companies experiencing continuous change. However, the combination of circumstances and priorities acting on any particular company is always unique. No one company can ever be said to exactly replicate another. It is for these reasons that different companies must approach the implementation of concurrent engineering by tailoring it to their own requirements.

Knowledge requirements for implementation

Companies do not therefore require panacea solutions for implementing concurrent engineering, but instead the knowledge and skills to understand better their own needs and potential solutions. The interaction between the various elements of the concurrent engineering solution may be complex and sometimes difficult to understand, but it is only through such an understanding that the best concurrent engineering implementation is achieved.

As each company looks at its individual requirements it will start to develop its own unique implementation of concurrent engineering. This diversity of implementation is to be welcomed, as it ensures that companies own their particular solution, they understand its nuances since they were responsible for developing it, and they are therefore capable of changing it in response to changing circumstances.

The challenge, of course, is for companies to have available a framework which aids them in understanding how the different sets of priorities affect their product introduction process and from there how to implement the best form of concurrent engineering. The first stage in this process is to understand how to classify the differences between companies; the second stage is to relate these differences to the types of concurrent engineering implemented.

Company variety

Chrysler and Boeing are two very different companies. The former develops a range of designs which it turns into millions of automobiles. Boeing, on the other hand, develops an extremely small number of designs and, relative to Chrysler, produces a small number of products – aeroplanes. However, are these differences significant in terms of the priorities being set for concurrent engineering? Both companies require extensive collaboration with suppliers, a very large development team, and long-term planning capabilities. Certainly both companies are held as examples of best practice in implementing concurrent engineering and in general terms they may be seen to have taken very similar approaches. The question then arises as to which are the common factors between Chrysler and Boeing that ensure that their two approaches are so similar.

A comparable situation occurs for our case study companies. Lucas Aerospace Actuation Division and Marconi Instruments are similar sized companies both producing sophisticated products with a high technological design content. In the two case studies significant reorganization of both companies occurred in order to develop product-focused teams. In that respect they followed very similar courses. However, in detail the final outcome is somewhat different. Lucas developed a balanced matrix structure where functional managers maintain expertise and programme managers coordinate project teams. In contrast, Marconi has developed product-focused divisions supported by functional service groupings. The differences in solution may be due to the fact that Marconi was moving away from the aerospace market which Lucas was remaining within. Alternatively it could have been due to other factors specified within the case studies. Whatever the differences between companies, and all the case studies do illustrate differences, it is understanding their effect on concurrent engineering that is crucial.

Explaining variety

Whilst it may be relatively easy to identify that differences occur between companies, it is not always so easy to identify what those differences imply, and what causes them. The product introduction process within any company is created to serve a variety of needs which in many instances conflict in priority and in how they are best satisfied. If a company such as Morris Cranes, discussed in Chapter 6, developed a product introduction process simply to satisfy its major business of one-off products, then it would become very inefficient when it came to progressing repeat orders. In contrast, if a company such as Measurement Technology, discussed in Chapter 5, developed a perfect product introduction process to suit its volume business, it would experience difficulties with low volume production or one-off customization. Both Morris Cranes and Measurement Technology must therefore have a clear view of all the likely priorities. In order to do this they must recognize how to respond to the influences acting on their particular circumstances.

A useful framework was developed in the 1980s by the management scientist Henry Mintzberg in his books *Structures in Fives* and *Mintzberg on Management* to describe differences between companies in terms of the forces they experience and the characteristics they adopt as a consequence. He depicted five categories of forces which act on all companies in a variety of combinations and strengths. These forces are described by the terms Direction, Efficiency, Concentration, Learning and Proficiency and are illustrated in Figure 2.1. Whilst no company ever experiences only one of these forces acting on its own, their effect is best described by considering that very situation. The pure characteristics which companies would be pulled towards as a consequence of the five pure forces are described by the five characteristics of Entrepreneurial, Machine, Diversified, Innovative and Professional. In practice companies become complex combinations of these five characteristics which interact in a variety of ways. The framework can be developed into a much more complex concept than described here to include the various relationships; however, for the purposes of this chapter the framework may be considered in its simplest form.

The five forces acting on all companies and their corresponding pure characteristics are described as follows:

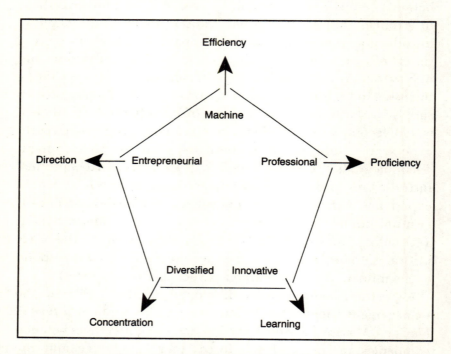

Figure 2.1   **Forces acting on companies and the corresponding characteristics they are directed towards (after Mintzberg)**

Direction This force is most evident when the organizational leader is strong and exerts a powerful influence to ensure that the organization moves in a specific direction. In practice this type of force is likely to exist within smaller companies as the focus of power is more likely to occur in one person. However, certain larger organizations where the senior management group is extremely focused can be seen to operate under a strong force of Direction. The result is to pull the organization towards an **Entrepreneurial** form, resulting in new ideas being introduced, implemented rapidly and with great conviction. The entrepreneur or small management group ensures a consistent outpouring of new ideas and products with a focused attention on ensuring success.

Whilst many smaller companies have a strong element of the Entrepreneurial form, a good example of this characteristic being exhibited in a larger company is Virgin Ltd. This company has a well demonstrated ability to surprise the marketplace with innovative ideas, as demonstrated by its moves into the airline, beverages and financial products businesses from its base in the music industry. The force for Direction is strongly evident within Virgin, led by its charismatic chief executive, Richard Branson.

Proficiency Organizations which pride themselves on their ability to perform their particular and often specialized tasks are responding to the force of Proficiency. These organizations often operate in environments where the retention of skills is critical to the survival of the company, and they are likely to be required to generate many of their skills in-house.

This force will pull the organization to take up the **Professional** form. This form is typified by a group of employees who take great satisfaction in demonstrating their skills. They recognize that their abilities are particularly tailored to their employer's business and consequently job stability is very prevalent. It is employees of this type of organization who suffer most when technology replaces their personal skills. Larger companies have, in the main, taken steps to reduce their dependence on a small group of technically skilled personnel by utilizing more advanced design and analysis tools. Within the manufacturing industry it is now the small specialist providers who are most likely to exhibit this form.

Learning This force represents the requirement for an organization to update its knowledge continuously in order that new products can be delivered to the customer. Whilst all companies clearly have to respond to this force, there are certain markets in which companies experience it to a very high degree. It is often found to influence those companies which compete on the basis of consistently introducing new products or product variants based on new knowledge or technology.

The form that companies are pulled towards in this circumstance is termed **Innovative**. The consumer electronics industry is a prime example of companies dominated by the Learning force. New ideas regarding product functionality are continuously being sought and to achieve this the organization is constantly learning both from its own activities and from monitoring those of other organizations. Sony Corporation is a well documented example of the type of company which has a strongly Innovative characteristic. New products are often regularly seen to include new electronic technology originating from within the company or adapted from external developments.

Concentration This force influences organizations to focus on their internal activities with little regard for the outside world, beyond satisfying their basic performance requirements. Managers at the middle level effectively take control of their activities, expecting little interference from outside once the terms of reference have been agreed.

The form that this force tends to create is termed **Diversified**. This is typified by companies which allow separate divisions to manage their own affairs whilst simply monitoring overall profitability and one or two other performance indicators. Many diversified conglomerates operate on this basis, probably the best known being Hanson plc. Individual companies are set stringent but straightforward targets to achieve and then left mainly to their own devices. Under these conditions the conglomerate's management structure is extremely simple and individual operating companies can be acquired and slotted into the business very easily. The downside of this characteristic is the lack of innovation and new ideas, leading to vulnerability in the longer term. Whilst in the short term benefits may be readily realizable, it is necessary to move away from this characteristic before too long.

Efficiency This last force is probably the one that is dominant in the environment of the 1990s for many companies. It is a force which drives companies to closely analyse their activities with the aim of achieving more for less. Where companies respond well to this force it results in procedures being developed to support best practice and detailed planning to ensure every action is carried out to support the overall objectives of the business.

The force of Efficiency naturally drives the business to take up a **Machine**-type characteristic where as far as possible uncertainty has been eliminated from all processes, and every action has a measure of performance associated with it. Many companies have experienced extreme examples of the force for Efficiency and have responded with dramatic change, often involving considerable downsizing. For those who have succeeded in passing through this phase and returning to healthy profitability (IBM is an excellent example) then the emphasis on becoming a Machine-type company will diminish to some extent. However, it is certain that the force for Efficiency will remain a strong influence on the activities of all companies for a long time to come.

This description of companies can help managers to understand the particular circumstances in which they find themselves and to determine a response. The framework based on the five forces acting on companies and the corresponding characteristics they are driven to assume is useful as a strategy-level tool. However, it contains within it all the elements that make up a company and not just those that concern the introduction of new products. In the following sections the framework is refined to provide a description which focuses exclusively on the theme of this chapter – the product introduction process.

The financial imperative

When Lee Iacocca set about restructuring Chrysler he was responding to a purely financial threat. He did not implement concurrent engineering because it was a radically new idea. He did it because he knew it was the only way to reverse Chrysler's deep financial malaise. Falling revenues from out-of-date products were sending Chrysler deeper and deeper into loss. Traditionally, the response to this situation would have been to boost revenue by extending sales through marketing initiatives and enhanced product features. Such an extension of product life-cycles was an approach that had certainly worked in the past.

However, the rules had changed in 1990. The option of extending product life-cycles no longer existed. The Japanese competition was developing totally new automobiles superior to anything Chrysler could achieve with existing models. Extending the life of current designs was not where the profit lay. Instead, it lay at the beginning of the life-cycle where a product brought to market early would reap the benefit of minimal competition.

Product life-cycles

The financial consequences of reducing product life-cycles is shown in Figure 2.2. It illustrates an idealized situation concerning income and expenditure associated with various stages in a product's life. Whilst no one product can be said ever to follow the idealized product life-cycle exactly, it is a useful concept to describe general trends.

During the initial development stage of any product there is relatively low expenditure due to the limited number of people involved on the project and the moderate investment in support tools. Since large-scale manufacturing facilities are rarely required during the initial design phases, it would be most unusual for any project to experience anything but a very small fraction of total development costs during this stage. Expenditure starts to accelerate as preparations are made for the product launch, the number of people involved in the project grows and investment is made in production facilities. Raw materials and

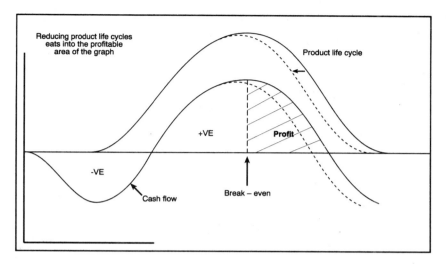

Figure 2.2 **The effect on profitability of reducing life-cycles**

component supplies are ordered to satisfy the initial production volumes. Expenditure rises rapidly as manufacturing facilities are brought on-stream, to be offset by the income stream generated as the product reaches the final customers. Once the product matures its profitability will begin to reduce as sales are seen to decline and lower cost competition enters the market.

This traditional and simplified description of the product life-cycle is usually accompanied by a description of initiatives available to stretch the life of the product, and thus maintain profitability for a prolonged period. These initiatives are primarily aimed at extending the right-hand portion of the product life-cycle graph after the peak in sales had been reached (or anticipated). However, in today's global marketplace these traditional activities are not necessarily sufficient in themselves to ensure financial viability. Global competition now means that product life-cycles are becoming much shorter, with obsolescence being less predictable. The product manufacturer has little control over when sales start to decline if so many competitors are available to enter the market.

It is not surprising, therefore, that attention has swung away from initiatives aimed at the later stages of the product life-cycle to those aimed at the earliest stages. If the demise of a product is controlled by the marketplace then the only option for a manufacturer aiming to extend the overall life-cycle is to hasten the product's birth. In financial terms this implies that, instead of relying on achieving a satisfactory return on investment through the extension of sales, the new priority is to start selling the product early.

Achieving break-even more quickly

Bringing products to the market early will only result in increased profitability if it carries a price and level of quality acceptable to the customer. If these requirements are met then sales will increase quickly and the critical financial point of break-even will be reached early. Concurrent engineering provides an approach to achieving this financial situation, but it does so in a combination of ways. The most obvious effect of concurrent engineering is indeed to reduce time to market. However, it additionally provides benefits through a reduction in product launch costs and through accelerated product sales. It is the combination of these three effects, illustrated in Figure 2.3, which provides such a financially attractive approach to product introduction.

The reduction in time to market ensures that break-even is

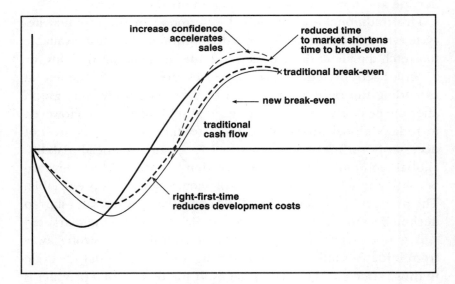

Figure 2.3 **Idealized options available to achieve break-even more quickly**

reached earlier provided the product is in a suitable condition to be launched on the market. However, since all aspects of the design are considered as early as possible, downstream problems typified by redesigns and delays are mainly eliminated. The knock-on effect of this approach is to provide greater confidence in the product introduction process and the commitment of funds to product launch. Based on this increased confidence, timely initiatives can be taken to promote and invest in the product to ensure a more rapid increase in sales volume.

Finally, the reduction in redesigns and delays eliminates a significant element of cost in the total development process. Experience has shown that, whilst an increase in engineering activity may be necessary at the very early development phases, this is easily repaid in reduced redesign and rectification at the vastly more costly stage just prior to product launch.

Investing in the early stages of product development

So rewarding is the opportunity provided by rapid product launch that companies are finding it beneficial actually to increase the amount of resource applied to the earliest stages of new product development. Figure 2.4 demonstrates a funda-

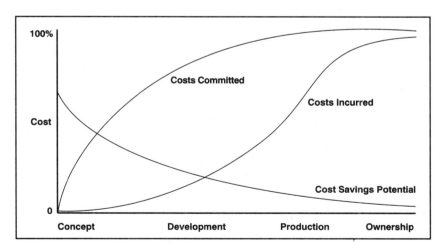

Figure 2.4 **A comparison of cost committed to cost incurred for new product introduction**

mental of all new product development. Whilst the initial stages of any project invariably require least expenditure, in contrast they define the vast majority of total life costs for the product.

It is at the concept stage of product development, where relatively few people are employed, consuming limited resources, that the most significant decisions are made. The majority of materials and manufacturing requirements are effectively defined at this stage, leaving opportunity for significant cost reduction later in the process to exist only in the margins. By ensuring that all downstream personnel are involved at the earliest possible stages of development, maximum savings can be achieved through a relatively low cost activity. Once this stage is past then the potential for reducing costs steadily diminishes – in contrast to the costs incurred which are accelerating rapidly.

Whilst it is generally accepted that all companies are experiencing pressure to deliver higher quality goods, at lower cost and more rapidly, the magnitude of this pressure varies significantly between companies. Certain industry sectors are experiencing intense competitive pressure from emerging competitors, whilst other sectors remain protected by high entry costs or particular geographical factors. As has previously been discussed, these pressures often follow cycles determined by the market or by the response of companies through change. This can be illustrated, for instance, by recent experience in the computer industry. IBM's decline and subsequent revival demonstrate how the consequence of reduced entry costs to the PC market can suddenly spawn a multitude of low cost competitors. IBM's response through a dramatic focusing of activity has resulted in a shift of its product strategy and a change in the competitive pressure that it is experiencing.

In addition to the pressures acting to produce better quality products more quickly and at lower cost, companies also experience other pressures which affect their new product development. In a world which is becoming more safety conscious and customers who are becoming more litigious, the internal procedures to be followed in developing new products increase in significance. Many companies' customers will require documentary proof not only that the products are to specification but also that the processes employed and procedures followed ensure that the products remain within that specification. Whilst the achievement of ISO 9000/1 may be sufficient for many manufacturing companies to demonstrate this ability, it is more complex for those companies which are contracted to design new products for their clients. In these instances the supplying customer must be able to demonstrate that it has the right people, tools, skills and knowledge available to complete the contract.

Another set of pressures acting on companies which affects their ability to develop new products may come from within the company itself. It is a common experience that a new managing director of a company will dramatically change methods of working to reflect his or her personal set of priorities. In some instances this new set of priorities may be totally in line with the basic objectives of concurrent engineering and will thus tend to emphasize this aspect of product introduction. However, it is not unusual for the new chief executive to focus on aspects of the business ignored by the previous incumbent. If the previous focus had been predominantly on concurrent engineering the new emphasis is likely to be elsewhere. Therefore under this type of circumstance the ability of the company to achieve concurrent engineering targets could as easily be diminished as it could be enhanced.

Whilst Marconi Instruments may describe in its case study the potential it sees in offering a product with increased functionality to a new market, another company such as Morris Cranes recognizes the need for increased customization. Yet another company such as D2D demonstrates that it is its ability to convince customers that it can progress a complex development project that is its order-winning criterion. One single concurrent engineering solution could never satisfy each of these three very simply differing requirements. The organizational structure, the people employed, the control systems put into place are just a selection of elements of the concurrent engineering solution that show significant differences between the three companies.

The ideal organizational structure for a company manufacturing mass produced products may include the design activity to be directed by a manufacturing engineer. In contrast, a manufacturer of safety-critical items would identify the need for increased emphasis on procedure development and may create a new function within the structure to ensure adherence to the newly developed procedures. The tools used in the design and manufacturing activities and the skills and expertise of the people involved would all vary, according to specific and individual needs.

Changing structures

Although many people take the concept of concurrent engineering to comprise product-based teams replacing old functional organizations, this is not the whole truth. It may undoubtedly be the case that in order to improve the product introduction process a more focused approach to activities is required. Under certain circumstances dedicated teams replacing functional structures is the best solution. But frequently a combination of the functional and project team approach is preferable to a purely project team approach, as many companies are beginning to discover.

A matrix-type organizational structure where functional managers and project managers overlap in their responsibilities is developing within many companies. The level of influence between these two management groups varies between companies, from situations where project managers are relatively lightweight in comparison to functional managers, to other situations where the project managers effectively form the senior management cadre within the company.

There are four basic management structures, illustrated in Figure 2.5, which experience has shown to be effective. In the three non-functional-type organizations illustrated, the relative strengths of the two management groups, functional and project, usually find a satisfactory balance to suit individual circumstances. The uptake of concurrent engineering concepts has probably been most dramatically illustrated by the shift of many companies away from their functional structure to one based on the matrix concept.

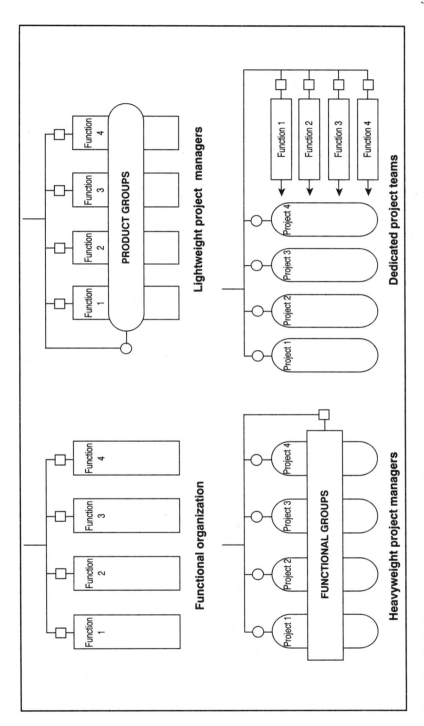

Functional organization

Lightweight project managers

Heavyweight project managers

Dedicated project teams

PRODUCT GROUPS

FUNCTIONAL GROUPS

Function 1 · Function 2 · Function 3 · Function 4

Project 1 · Project 2 · Project 3 · Project 4

Figure 2.5 **The four basic structures for management**

However, experience has shown that companies which have moved directly from a functional organization to a project-based structure have, over time, actively moved to reintroduce some of the elements of functional structure. Frequently the requirement for this is to ensure long-term retention of skilled people through the provision of an identifiable location within their peer group. The undoubted benefits of pure project-based structures bring with them certain long-term deficiencies which can only be eliminated by some form of management matrix which includes a functional element.

Changing tools

The application of computerized tools is often seen as central to a company's implementation of concurrent engineering. The take up of CAE (computer aided engineering) tools is strong within all of the case study companies – although they do not all emphasize this issue. In addition, EDM (engineering data management) is emphasized by those companies associated with the computer industry – IBM and D2D. The sophistication of CAE tools employed depends on specific company requirements. For some it is sufficient that they have available dedicated analysis routines. For others a sophisticated three-dimensional modelling capability to support complex stress analysis is required.

However, it is interesting to note that some of the most effective tools associated with a concurrent engineering implementation are non-computerized and relatively simple in operation. The most important of these is the tool used to translate customer requirements into product descriptions – quality function deployment. This tool, also known as the House of Quality, originated in Mitsubishi's Kobe shipyard in 1972. It was refined to become the Koe-Kikaku or 'voice-planning' tables which give a graphical method to convert the voice of the customer into product definitions. It was first reported to the western world in an article by John Hauser and Don Clausing in 1988 in the *Harvard Business Review*. A typical example is shown in Figure 2.6.

The strength of this tool lies in its ability to form a communication link between groups such as design engineers,

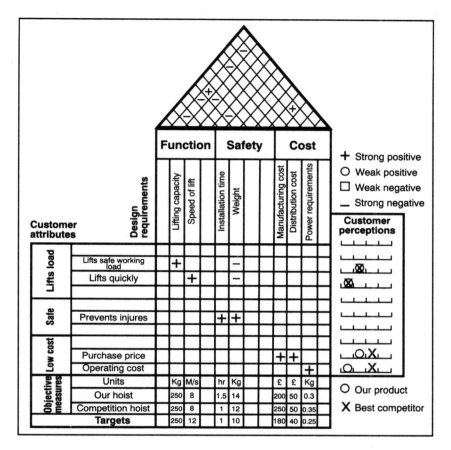

	Function		Safety		Cost			

Legend:
+ Strong positive
O Weak positive
□ Weak negative
— Strong negative

Customer attributes	Design requirements	Lifting capacity	Speed of lift	Installation time	Weight	Manufacturing cost	Distribution cost	Power requirements	Customer perceptions
Lifts load — Lifts safe working load		+			—				⊠
Lifts quickly			+		—				⊠
Safe — Prevents injures				+	+				
Low cost — Purchase price						+	+		O X
Operating cost								+	O X
Objective measures — Units		Kg	M/s	hr	Kg	£	£	Kg	O Our product
Our hoist		250	8	1.5	14	200	50	0.3	X Best competitor
Competition hoist		250	8	1	12	250	50	0.35	
Targets		250	12	1	10	180	40	0.25	

Figure 2.6   **A typical House of Quality**

manufacturing engineers and marketing personnel. It takes as its starting point a detailed list of customer attributes which define product features in the language of the customer. The engineering characteristics of the product are then created by comparing what the design engineers are able to achieve with its effect on the overall product design in terms of satisfying the customer. Thus each feature of the product is gradually specified in engineering terms, whilst trade-offs between the various design options can be carried out using an established set of metrics.

Changing people, processes and controls

Ford has determined that in the environment of global manufacturing the only competitive weapon that can be protected from being replicated are its employees. It therefore embarked on a programme of continuous education in the 1980s which has now grown into a transatlantic initiative whereby national universities provide training for hundreds of Ford employees up to postgraduate degree level every year. This commitment to enhancing the skills and knowledge of employees would be difficult for any but the largest company to imitate. However, it is undoubtedly the case that a fundamental element of any successful concurrent engineering implementation concerns continuous training of the people employed. In a world where change is ongoing, only those people trained beyond the level of their current activity can be relaxed about the consequences.

Undoubtedly, changing the attitudes of people, the processes that they follow and the controls that are utilized has one of the largest effects on the introduction of concurrent engineering. Changing the way individuals work is always difficult, especially so if the general perception is that there is little danger in maintaining the *status quo*. Inertia, present in all companies but especially in larger ones, almost always acts against the introduction of new ideas where the likely result is simply an improvement to an already satisfactory position. It is possible to achieve change through strong directive management and in some situations it is undoubtedly the most effective solution. However, the long-term aim must be to ensure that all employees understand why there is a need to change and are prepared for it. Ford is driving hard in this direction through its programme of continuous learning. Each company must find its own way to achieve a similar result.

A framework for concurrent engineering

It has been seen in previous sections that a framework can be used to describe the pressures acting on a company and the forms it will take up in response. In addition, the various elements of a product introduction system have been discussed in terms of changes that may be required to implement concurrent engineering. These concepts are now brought together to develop a framework for concurrent engineering.

The process of developing a new product starts with an initial realization of need and finishes with all the essential actions, knowledge and information required to launch the product. This is a definition of the product introduction process – see Figure 2.7. Within this definition it is assumed that all necessary information related to manufacturing, distribution, after-sales service, maintenance, recycling, disposal, etc. which is relevant to the product's design has been incorporated.

The framework for concurrent engineering is based on the concept that the objective of concurrent engineering is to improve the product introduction process. Just as the previously presented framework in Figure 2.1 consisted of pressures acting on companies to drive them to take up a combination of characteristics, so the framework for concurrent engineering has the same underlying philosophy. It is shown in Figure 2.8 and includes the five forces that can act on a product introduction process: those of Efficiency, Focus, Proficiency, Radical Innovation and Incremental Change.

These pressures acting on the product introduction process differ slightly from those seen to act on the company as a whole. This reflects the fact that not all external company pressures act in their pure form on the product introduction process. Whilst the pressure for Efficiency and Proficiency can be identified at both the company and product introduction process levels, other pressures are modified slightly. Thus a company-level pressure for Learning will be reflected as a pressure for Radical Innovation on the product introduction process. Similarly, the company pressure of Concentration will be seen as primarily responsible for instigating a pressure of Incremental Change on the product introduction process. Finally, the pressure for Focus acting on the product introduction process is a direct result of the company-level pressure for Direction.

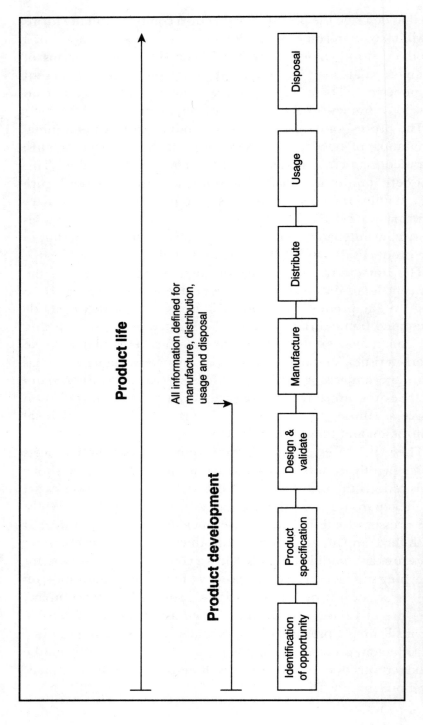

Figure 2.7 The product introduction process

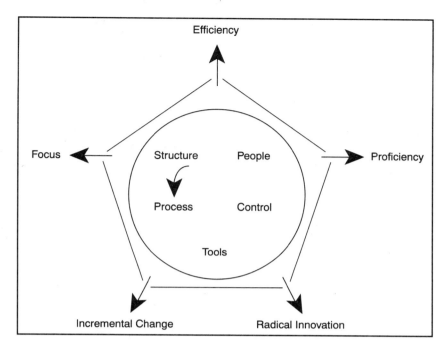

Figure 2.8 **The concurrent engineering framework**

The characteristics of the product introduction process are then defined to comprise elements of Structure, Process, Control, People and Tools. This is a commonly accepted set of elements to describe any organizational system. The one difference in approach to the previously described framework is that the individual forces do not relate to any one specific element, but to all the five elements which are shown within a rotating wheel. This represents the concept, previously described, whereby a company will move on from one priority to the next, once sufficient improvement has been seen to occur.

The product introduction process forces

The five forces acting on a company's product introduction process relate to the pressures it experiences to introduce new products. It is a rare company that does not experience the pressure for Efficiency, and similarly most companies have experienced a pressure for Focus brought by a strong leader.

Other pressures are usually more specific to certain company types. Some companies aim for Proficiency by demonstrating their capabilities, others look for rapid product change through Radical Innovation, and finally other companies look for steady enhancement through Incremental Change.

Efficiency   This is the force that is driving companies to develop their new products in a shorter time than before, at lower cost and with increased quality. It is the force which tends to dominate in today's market environment and as such it is placed at the top of the diagram. It can be assumed to contain the driving forces for the introduction of concurrent engineering, which is then influenced by the following four forces.

Focus   This force relates to the application of strong leadership to drive the activities of the product introduction process in a specific direction. In that respect it can be supportive of any of the other forces, tending to reinforce rather than drive in its own particular direction. In the ideal world the force of Focus should drive the product introduction process in the direction that is most beneficial to the company's future. However, the combination of strength and wisdom is not particularly common within individuals and the force of Focus may not always act in the best direction.

Proficiency   This force acting on the product introduction process acts to drive a company to demonstrate to customers its capabilities in developing new products. Those companies involved in winning large contracts for capital goods design and manufacture, or in providing manufacturing services, must demonstrate their ability to perform. Customers will be seeking to identify demonstrable Proficiency prior to awarding any contract.

Radical Innovation   For some companies competitive pressures may force the development of radically new products. This is a natural tendency of rapidly developing markets based on new technologies, typified by the knowledge-based industries and biotechnology. However, more traditional industries may have to respond to this force when a new and strong competitor arrives in the marketplace.

Incremental Change Many companies experience the force for regular updating of their product range. The global market is exerting pressure for this activity to occur more quickly and more often. However, in essence Incremental Change is building new products based on past experience. Many manufacturing sectors require prior knowledge to be incorporated to a very large extent in new products in order that both internal and external confidence is maintained.

It will be immediately apparent that most, if not all, of these forces will be acting on a product introduction process at any given time. The development of any new product is likely to be driven by both Radical Innovation and Incremental Change, since previous experience will usually be supplemented by new technology in specific areas. There will be the pressure for Efficiency in minimizing costs and time to market, whilst at the same time there will be a need to demonstrate ability to complete the project, i.e. Proficiency. Finally, there will be leadership of varying strength providing the Focus for all activities.

The product introduction process characteristics

The characteristics of the product introduction process are taken to involve the five elements Structure, Process, Control, People and Tools. This set of elements is well accepted as comprising all aspects that must be considered in developing any organization. It is therefore able to describe fully any design of product introduction process that can be envisaged. The benefit of this approach is evident when the detail of each characteristic is described. Rather than having to define individual product introduction processes in all their complexity, this approach allows for individual elements to be defined in order to create a unique combination.

Structure At one extreme of Structure is the traditional hierarchical pyramid based on a functional organization where people are grouped by the type of operation they perform. The other extreme consists of product- or programme-focused groupings which typically comprise multifunctional teams concentrating on a specific product range. Most companies operate with a matrix-type organization which, depending on circumstances, will exist somewhere between these two extremes.

Process   The two extremes of Process can be considered as, on the one hand, sequential processes where each action follows on from the previous and, on the other hand, parallel where all tasks are carried out at the same time. In practice it is rare to find companies which operate at either of the two extremes, and most conduct some operations in parallel and some sequentially. Typical of this is the situation where formal milestones are imposed on the product introduction process. Between the sequential milestones parallel operations take place.

Control   The way in which projects are planned and measured in terms of milestones, deliverables, costs etc. is covered by this characteristic. It also includes the mechanisms in place to encourage other initiatives such as continuous improvement and total quality. Formal project planning and monitoring systems can range from fully automated systems to simple manual approaches, depending on circumstances.

People   This element is concerned with the roles that people take within the product introduction process and the level of knowledge that they require. In traditional organizations the role of individuals within a product introduction process could often take the form of a specialist activity working in an individual manner. This type of working has changed significantly in many companies since there is a requirement for people with more generalist skills capable of taking on a range of roles and working closely with a variety of people in a team structure.

Tools   This element covers all the range of tools and techniques likely to be employed within the product introduction process. It includes CAE tools comprising draughting, modelling and analysis; IT tools such as engineering data management, electronic communications, videoconferencing, dedicated data lines etc.; and various quality tools. It includes sources of information such as design procedures, product data, customer information, and feedback from projects. Finally, it includes manufacturing equipment, which may have a major impact on the product introduction process through new process capabilities.

Using the framework

We shall see in the case studies that follow that each of the companies can be related to the concurrent engineering framework. IBM Havant, for example, describes a situation where there is a very strong pressure for Efficiency with a combination of Radical Innovation and Incremental Change. The company responds by focusing activity on the Tools element of the product introduction process and to some extent on the Process element. This is not to say that it is ignoring the other elements of the product introduction process, it is simply that they have already been satisfactorily addressed under previous initiatives.

In a similar situation Measurement Technology is seen to concentrate on changing the Structure element of the product introduction process. It operates under the same pressures as IBM Havant but its response is very different. The previous functional organization is converted to include business groupings in order to provide greater direction to its activities. The driving forces for IBM Havant and Measurement Technology are very similar, but the resulting actions are very different. However, both situations can be described relatively easily in terms of the concurrent engineering framework. Similarly, the responses of all the other case study companies can be described by the framework. This is indeed the primary function of the framework, to accommodate all situations that companies may experience in order that a clear and precise description can be created.

In all fields of life it is only through this activity of placing experience within a framework that understanding develops. The application of concurrent engineering is no exception. There is naturally some simplification that occurs in the process of using the framework. It can never provide all of the detail that any individual would need. However, this simplification is vastly outweighed by the understanding that comes with its use. The framework provides practising managers with an understanding of where they should be going and why. It is then up to the individual to provide detail based on knowledge and understanding of their own specific situation in order to create a suitable concurrent engineering environment.

The case studies

The case studies which follow are presented mainly in pairs, although in one instance three are presented together. The objective of this approach to presentation is to contrast and compare the different approaches taken by the various companies to implementing concurrent engineering. It will be seen that where there are similar pressures the response can take very different forms. The intention is for the reader to recognize that none of the responses is optimal, each simply represents a reaction to particular circumstances.

The concurrent engineering framework is used to describe the case studies and place the activities in context. However, it should not be forgotten that the framework is there to aid understanding, it is not a predictive tool. It will help at the level of identifying issues to consider, but it is then up to the individual to determine the detail.

Bibliography

Backhouse, C.J., Brookes, N.J. and Burns, N.D. (1995) 'Contingent Approach to the Application of Concurrent Engineering to the Product Introduction Process', *Proceedings of the IEE, Special Issue on Manufacturing Engineering*, **142**, (5), September.

Hauser, J.R. and Clausing, D. (1988) 'The House of Quality', *Harvard Business Review*, May–June.

Mintzberg, H. (1983) *Structures in Fives: Designing Effective*

Organisations, London: Free Press.

Mintzberg, H. (1989) *Mintzberg on Management: Inside our Strange World of Organisations*, London: Free Press.

3
Structures and Processes

Marconi Instruments Ltd and Lucas Aerospace Actuation Division

This first chapter introducing contrasting case studies of concurrent engineering concerns two similar-sized companies which have followed what many people accept as the classic implementation route. The companies, Marconi Instruments Ltd and Lucas Aerospace Actuation Division, have placed the highest priority on creating organizational structures to focus more closely on new product introduction and on refining their new product processes. Both companies have developed matrix-type structures which retain some elements of a functional organization but include strong programme management activities, either encompassing or operating across functions. In addition, they have both formalized all the stages that any new product development must follow.

In terms of the concurrent engineering framework introduced in Chapter 2 (and illustrated in Figure 3.1), the two companies can be seen to have been strongly influenced by the forces of Efficiency and Radical Innovation as their traditional marketplace diminished and they had to realign their activities. In addition, both companies experienced the reinforcing of Focus with a strong commitment to implement concurrent engineering being present in both senior management groups. The priorities for change were again similar in that both companies focused on the characteristics described by the elements of Structure and Process.

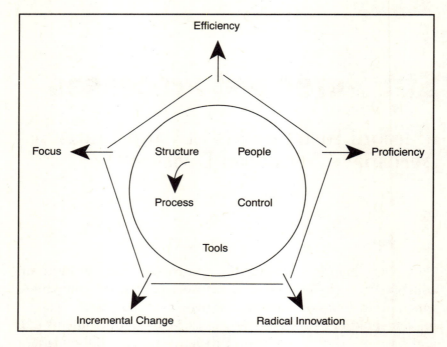

Figure 3.1   **The concurrent engineering framework**

The pressure for change

These two case studies concern companies that experienced a
dramatic change in market conditions as a result of the decline in
orders from the military sector. As part of radical realignment of
activities, both companies recognized in the late 1980s that they
would have to shift their focus from relying on the highly specific
defence industry products and extend their activities to within
the civilian market. The new market conditions experienced by
Marconi Instruments and Lucas Aerospace Actuation Division
were very similar. Whilst pressures remained in the cost and
quality areas, a significant new pressure was experienced which
concerned the provision of new products on a timescale which
neither company had experienced previously. The necessity for
radical innovation of their products was fundamental to the
ability of both companies to survive within the new market
sectors.

Marconi Instruments probably experienced the greatest shift in emphasis as it moved away from purely military applications to a general market for its communications test sets. Lucas Aerospace Actuation Division also moved away from military applications but into the relatively more tightly defined market (when compared to Marconi's experience) of civilian aerospace. Whilst this disparity in emphasis resulted in different concurrent engineering solutions, it undoubtedly resulted in a similar level of effort and change within the two companies concerned.

The change

The first difference to note in the implementation of concurrent engineering is the route adopted. The approach taken by Lucas Aerospace Actuation Division was to create task-forces which would be responsible for analysing the current situation and developing a plan for the conversion of the complete business unit to the concurrent engineering approach. The next stage was then to implement a demonstrator whereby one specific product development, the thrust reverser engine actuation system, was used to demonstrate the new way of working.

In one respect the approach of Marconi Instruments was similar, since it too demonstrated concurrent engineering through one individual development – the microwave test set. However, whilst the motivation for Marconi Instruments was to convince the rest of the business that concurrent engineering was a worthwhile way of working, it was slightly different for Lucas Aerospace Actuation Division. Its motivation was to measure how effective its proposals would be in order to identify improvements for the forthcoming, and inevitable, reorganization.

The second issue to note is the differing priority placed by Marconi Instruments and Lucas Aerospace Actuation Division on the relationship between structure and process. In the case of Marconi Instruments the priority was to identify more market-oriented business units, which resulted in new product processes being introduced at the same time as the reorganization. In contrast, Lucas Aerospace Actuation Division focused more closely on the new product process, which was itself redesigned prior to the introduction of the new company structure. This difference in approach again illustrates the fact that particular circumstances are responsible for determining the correct route through any concurrent engineering implementation and that there is no single panacea approach.

It is also interesting to note that the final organizational structures which the two companies developed are different in some significant ways. Marconi Instruments chose to create financially responsible business units for each market sector, supported by core service units. In contrast, Lucas Aerospace Actuation Division retained a structure which emphasizes the various engineering functions, with cross-functional programme managers providing the coordination and control of new product development.

Both these two company structures are correct since they reflect the particular set of circumstances in which each company found itself. Marconi Instruments has a fairly diverse range of markets, each of which has specific requirements. In contrast, Lucas Aerospace Actuation Division has a more focused market with less need for specific business units. It is therefore to be expected, and is certainly not surprising, that different forms of concurrent engineering resulted.

In terms of new product processes both companies recognized the need for a formal process which would be followed for all new product development. This was clearly a response to the need for more reliability in the product development process, with greater emphasis being placed on guaranteed new product launch dates. Both companies identified stages within the process, with formal review mechanisms and senior management approval required for progression between stages.

Finally, the two case studies demonstrate a similarity in approach to the application of new tools to support product development. In both cases, the change to the organizational structure and the process of developing new products was defined before any major change in the availability of support tools was considered. This approach reflects accepted wisdom that investing in sophisticated tools is rarely effective unless a prior business analysis and process simplification have been carried out beforehand.

On-Time Product Design

Marconi Instruments Ltd

Tim Pegg

Company background

Marconi Instruments Ltd is a manufacturer of test and measurement equipment for the communications industry. The company, established nearly 60 years ago, began by making audio products. It now employs 1100 people worldwide, including 950 on three sites in the UK. Turnover is approximately £70 million, of which 80 per cent comes from exports. It provides products and solutions in four main areas: radio communications, microwave, telecommunications and printed circuit board test.

Competition is mainly from very large multinationals with well-established customer links and reputations. In this respect there is a need to provide the customer with sufficient incentive to choose a Marconi Instruments' product over that of its bigger competitors. This provides a significant challenge for Marconi in overcoming any perceived 'risk aversion' by customers – it has often been said that no-one gets fired for buying a piece of Hewlett-Packard equipment!

The company is primarily a design-to-forecast organization in that it designs new products based on a speculative prediction of long-term market requirements. It enjoys relationships with its customers that allow informal exchanges of information in areas where mutually beneficial technology developments might lie.

This is a process which Marconi sees as offering significant benefits for both itself and its customers and which it intends to develop further. The volume of units manufactured for a given product type tends to be small, of the order of hundreds or a few thousand. In addition, the market environment dictates small batch sizes and a complex mix of products.

In the late 1980s the company underwent a thorough rethink of its strategy regarding core technologies. This led to decisions concerning which items should be sourced outside and which should be brought back into the company. Marconi now buys in sheet metal fabrications and cables, but retains as key to its business the manufacture of printed circuit board (PCB) assemblies and microwave components.

The 6200 series Microwave Test Set

The field of microwave measurements has historically been dominated by military and aerospace requirements. This has led to the development of expensive and sophisticated equipment, often providing a number of obscure facilities not required by the majority of users. The decline in the military market and increase in microwave communications has led a number of companies, including Marconi Instruments, to rethink their commercial strategy. Users now require equipment to provide a cost-effective solution to their particular measurement needs and are not prepared to pay for redundant features.

In 1988 Marconi Instruments examined its best-selling product, the 2955 Radio Communications Test Set, and found that its key advantage at the time of its launch was the unique combination of functionality that it contained. It was the fact that the 2955 had been extremely successful in its particular market sector that stimulated Marconi to see if a multifunctional test set could be successful in a different market sector. In consequence, the 6200 series Microwave Test Set (MTS) was conceived as a product containing not only the most commonly found items of microwave test equipment, i.e. a synthesized sweep generator, a scalar analyser, a power meter and a frequency counter, but also a programmable voltage and current source, and fault location software. Packaging these elements into a cost-effective, single, small portable unit was a new concept in microwave test equipment.

A strategic decision to invest heavily in microwave technology provided the background against which the 6200 series MTS was introduced. It was the first new microwave product for a significant time and provided the initial opportunity to see a pay-back on the strategic technology initiative through increasing Marconi's microwave test equipment market share. It was recognized that to provide this return and reach customers at the right time, lead-time for this product would have to be significantly shorter than the four or five years that such a complex product would normally have required. The 6200 series MTS had a lot of expectations to live up to!

The new approach

In 1988 Marconi was already beginning to consider moving away from a functional structure towards one that was more product-focused and the development of the 6200 series MTS provided the ideal opportunity to prove the new approach. Proposals to develop new products using multidisciplinary teams received a sympathetic hearing from senior management and support was more easily obtained. Elements of concurrent engineering had begun to emerge in the company in the late 1980s, but the start of the development of the 6200 series MTS in late 1989 provided the first specific focus for the project-based approach.

In order to achieve its targets, it was recognized that the development of the 6200 series MTS would have to overcome the following three key challenges:

Up-front specification It had been traditional within Marconi Instruments that product specifications would rarely be finalized until initial prototypes had been produced. This resulted in prototypes being completed early in the process, but clearly led to delays and cost overruns during later stages. With pressures on time becoming so dominant it was clear that downstream problem rectification simply could not be afforded, and that right-first-time design was a necessity. This was achieved by insisting that no design work would start until a detailed and precise specification of the requirements of the 6200 series MTS had been finalized. This comprised both the product requirements specification and the man–machine interface specification.

The time taken to complete the specification stage was not inconsiderable and, it cannot be denied, led to a certain amount of unease on the part of management and engineers alike. Extending the time taken to develop the specification presupposed a consequential reduction in the, traditionally problematic, development time. Lacking previous experience that this was actually the case instilled a significant feeling of risk.

The decision was fully vindicated by the minimization of downstream problems as the project progressed. A gratifying example of this was the design of the analogue and digital PCBs, the 'heart' of the instrument. The two PCBs are large, complex multilayer boards with a mixture of double-sided surface mount and leaded components. They function as a pair and so it was vitally important that both were relatively 'fault free' when the two boards were connected together. Four and a half months after the start of the project the PCB artwork technician began the layout of the board. Three months later the design engineers had populated boards back in their hands to begin testing. Only six minor faults had to be corrected to get them to function. These were equally split between the two boards and luckily half were the responsibility of the design engineers and half that of the artwork technicians.

In contrast to previous experience the prototype boards were produced later in the project, but subsequent time and resource required for design changes and fault rectification were dramatically reduced.

Communications   The previous functional organization of Marconi Instruments had dictated an environment where communication between various groups had supported the development of high quality products but had not supported short development timescales. A traditional over-the-wall mentality had pervaded product development with the consequential communications problems at the interfaces.

In order to facilitate communication, the project was undertaken by a full-time co-located multidisciplinary team which was supported by a number of part-time team members. All of the people involved in the project, about forty in all, were brought together at the project inception for a briefing session. Many of the participants would not be asked to contribute significantly in terms of workload until several months later. However, their input at this initial stage was both substantial and important for future success.

A significant level of impromptu discussion resulted from the close proximity of team members. Good communication resulted in effective decisions being made at the earliest stages in the crucial trade-offs between commercial issues such as cost, size and performance. In addition, proper attention could be paid to production aspects such as manufacturability, testability and serviceability. These activities took place prior to development of the first prototype, the exact reverse of previous practice.

Weekly progress meetings were held for the core team, with the updated progress report being circulated to all team members (full and part time) to ensure that everyone was kept informed of the project's progress. Because of co-location, informal and impromptu discussions and meetings occurred continuously. Only when serious problems occurred was it necessary to call an emergency review.

Reuse of existing technology   It was a conscious design decision that existing technology should be employed in order to save on design resource and crucially to reduce time to market. Previous practice had been to allow engineers to develop completely new component designs for each new product. Whilst this undoubtedly led to the highest technical specification being attained, it no longer made commercial sense. The reuse of well proven components would both reduce development time and ensure that many of the manufacturing problems typical of a new product would be eliminated from day one.

Three significant areas utilized existing technology, thus allowing the design engineers to concentrate their innovative abilities in a narrower field of activity. The 6200 series MTS case was an adaptation of the 2955 Radio Communications Test Set, the power supply was derived from the 6311 Programmable Sweep Generator and the Fractional N Synthesis System was easily adapted to integrate into the 6200 series MTS. In consequence, the design engineers could then concentrate on developing a highly successful colour display and employing new, smaller and cheaper components within the 6200 series MTS. These new components were, in fact, the consequence of several years' research by Marconi Instruments in specific enabling technologies. Thus the design team concentrated on those aspects of the project which were innovative and which would be significant differentiators of the product in the market.

The application of these three basic mechanisms was successful in assisting in a very favourable product launch for the 6200 series MTS in 1991. Not only was it introduced in a lead-time of less than two years (a reduction of 50 per cent on what would have been expected if a more traditional approach had been used), but it has also tripled the turnover of Marconi Instruments' microwave product range in less than three years. It was sufficiently radical in its functionality to ensure that it was only in 1994 that competitors introduced a similar product.

The new company structure

The success of the 6200 series MTS clearly showed that a concurrent engineering philosophy could work for a particular project. The challenge for Marconi Instruments in 1991 was therefore to build on this experience to provide an approach for implementing new products throughout the company. This was achieved by initially reorganizing the company into a series of business units with profit-and-loss accountability for the products that they provided.

The business units contained most, but not all, of the design engineering and marketing resource required for the product range for which they were responsible. Additional supporting resources were available via a number of core services most suitably maintained as centres of expertise. These consisted of manufacturing, purchasing, sales, finance, PCB artwork, technical publications, drawing office and systems engineering – see Figure 3.2. In practice, each core support area comprised between five and ten employees on a company-wide basis and could therefore cope with the changing demand from various business units. Exceptions occurred in sales, finance and manufacturing which consisted of larger groups of employees.

Naturally, both the business units and the core service groups experienced peaks and troughs in demand. The company policy

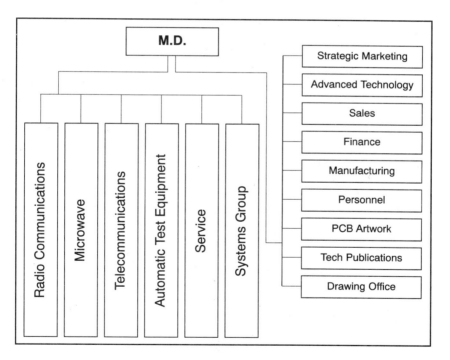

Figure 3.2 **Company structure illustrating the relationship of business units to core services (UK only)**

was to alleviate this problem through a flexible and cooperative attitude to secondments. In this instance the overall goals of the company to maintain a highly skilled workforce took priority over immediate business unit objectives. Secondments also provide a mechanism for career development for the engineers concerned.

Marconi Instruments was well aware of the potential hazards, as well as the benefits, of operating in a product-focused fashion. Being focused around product lines rather than market segments, the same customer may interface with several parts of the company leading to the development of 'overlaps' and 'holes'. One of the key approaches to avoiding this was the creation of three integrating mechanisms working through a series of committees to provide an overall view on technology and product development.

- The **Strategic Marketing Group** comprises a number of marketeers having a brief outside the business units. Their role is specifically to identify new and/or declining markets and to spot holes between product group portfolios.
- The **Product Approval Board** is a subgroup of the senior management team and makes strategic decisions on which products should be developed based on recommendations of the Strategic Marketing Group and the business units. This subgroup meets quarterly as part of the company's formal procedures.
- The **Advanced Technology Group** comprises approximately five engineers working in areas of technology key to those future products identified by the Product Approval Board. Activity in this area may be partially contracted out to external sources of expertise, with knowledge brought back into the company as developments occur. The final product may not have been precisely identified at this stage.

The team structures

Within the framework of the business units, full-time multidisciplinary teams have been established for new product development. The full-time team is co-located and its membership comprises design, production and test systems engineers and a technical author. All, apart from the design engineers, are seconded to the team using a 'matrix management' approach. In addition, where a particular business unit is short of engineering resource for a particular project, this may be seconded in from another business unit. This is naturally dependent on the availability of resource.

Crucial to the success of the secondment approach is a consistent attitude towards standards which must be common across all business units. The responsibility for this lies with coordinating engineering working parties who determine standards and with functional heads who ensure compliance for their disciplines. The functional heads are additionally responsible for longer-term planning and resource balancing, which involves the coordinated transfer of people between the business units and/or projects. The day-to-day work planning remains the responsibility of the project leader. Finally, it should also be noted that groups of engineers become established as 'centres of excellence'. Thus the microwave component design engineers are a 'centre of excellence' for the company and design microwave components for other business units.

The project leader is appointed with responsibility for development time, cost and technical decisions. On larger projects the technical decisions may be the responsibility of a 'technical' project leader. Additionally a shadow project leader is appointed at the same organizational level as the project leader, but from a different business unit, to act as a reviewer for the project.

Co-location is considerably eased by the relatively small number of total people involved. Full-time co-locatees are often not only sitting in their team but are also next door to their original sections. Part-time support is provided by PCB design, draughting, prototyping, production liaison, purchasing and the service business unit – see Figure 3.3.

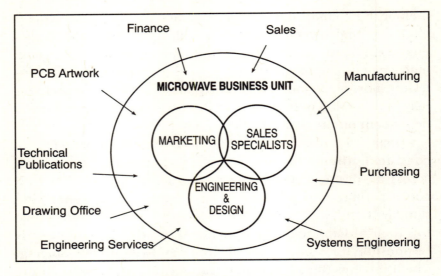

Figure 3.3   **A typical business unit receiving input from core services**

The new product introduction process

Corporate procedures detailing the way in which new products are introduced are not new to Marconi Instruments. The existing internal process control mechanisms and the introduction of external quality standards (ISO 9000) has required that proper procedures be maintained. When the teamworking practices tried out during the 6200 series MTS development were introduced on a widespread basis changes were required. This did not in fact pose any major difficulty as, in the main, only additions were required. The most significant change concerned the focus on gaining a very clear specification of the whole project before any work commenced.

The new process also placed a great deal of emphasis on documentary evidence of reaching key milestones and on assessing the project via peer review. The process is formally divided into three stages, comprising product investigation, product development and release for manufacture – see Figure 3.4. Although the end of each of these phases is clearly marked by a deliverable, there is also a requirement for a series of milestones with related deliverables within the individual stages. This

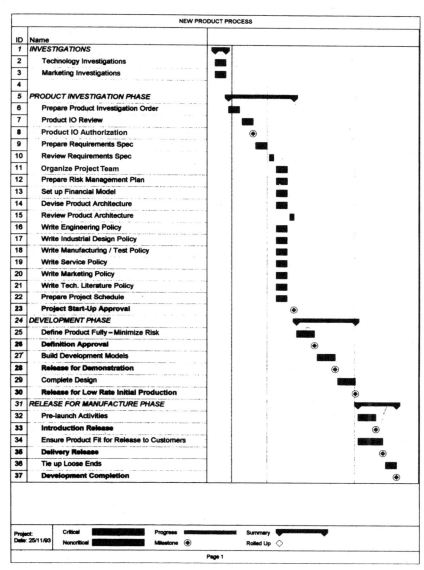

Figure 3.4 **Marconi Instruments' *New Product Process Pocket Guide* illustrates a typical project Gantt chart**

approach of identifying clear stages in the process remains compatible with the objective of parallel development, although it clearly introduces some element of sequential activity. Overlap can exist between phases, but this is dealt with 'by exception'.

Thus the late availability of one phase's deliverable may not delay the initial activities within the next phase.

Project start-up approval

One of the most important deliverables is the project start-up approval document. This marks the end of the product investigation phase and refers to:

- a project plan
- a financial model
- a technical requirement specification
- an assessment of areas of high risk with contingency plans
- a list of people to be responsible for the project assigned by name
- policies defining the requirements of downstream activities such as manufacture, test and service.

Only when this document has been formally signed can development work on the project commence. This original document has been streamlined and focused since its introduction, especially in terms of the approval signatures required – if a person has been empowered to sign a document there is no point in their manager also signing it. Any gaps in the required information would result in the project proposal not being considered. These changes have gone a long way towards removing people's initial reservations.

Implementing change

Initial resistance was felt from some quarters towards the new product introduction process (NPIP) when it was introduced. Some felt that it was being forced on them by the Microwave Business Unit. However, others, who had witnessed the benefits and successes of the 6200 series MTS development, acted as champions within their own teams. Criticisms of the NPIP document were due to it being based on deliverables with a consequential large volume of paperwork. This was a result of the NPIP being intended to be explanatory and helpful, but the thickness can put some people off. The document is currently being simplified to replace text with flowcharts and diagrams. This is an ongoing process since the NPIP is continuously evolving and is not a 'do once, last for years' process.

Future introduction of support tools

A conscious decision at the outset of the 6200 series MTS project was to concentrate on changing the way in which products were introduced and not to change the tools used. This approach was seen as providing maximum return quickly, in contrast to the dangers inherent in applying new technology to out-of-date procedures.

By 1993, however, the changes to structure and process were well under way and attention turned to using CAE within the product development process. Purchase plans were finalized for a PCB development package to compress product development lead-times even further. Campaigns to install the 'voice of the customer' into product development were also commenced. The marketing expertise within the company was strengthened by recruitment and internal secondment to put more effort 'up front' in identifying customer needs. Quality function deployment was used at the outset of some projects to map customer needs against product functionality. It was the aim to have all engineers make at least two customer visits every year. This may be up front when specifying a new product, or it may be whilst carrying out beta site trials on a product just before it is launched.

Additional techniques such as FMEA were also investigated, although there were no immediate plans to introduce this as Marconi Instruments was very conscious not to 'flood the system' with change but to ensure that key changes took hold.

Conclusion

The form of concurrent engineering used at Marconi Instruments clearly has its roots in the way in which the successful pilot on the 6200 series MTS was conducted. This acted as a template for all future product development and its 'best practice' was applied, with some modification, across the whole of the company.

The key lessons learnt in the implementation were:

- to use a single project as a demonstrator before implementing company-wide change;
- to accept the perceived wisdom that the development of a complete specification does significantly reduce overall product development time;
- to introduce processes which ensure that long-term company objectives retain their influence on the project-based structures of the business units;
- to change the process of developing new products before attempting to introduce new technological tools.

Undoubtedly the success in implementing concurrent engineering can be attributed to the willingness of all employees to change. Whilst this was made easier by the small size of Marconi Instruments, which has always enjoyed a friendly and flexible relationship with its employees, it is also true for many larger companies. The important factor is to ensure that the changes are well thought out and fully supported by senior management.

Finally, it must be stated that the process of change cannot be underestimated. It took over five years to introduce concurrent engineering to Marconi Instruments, and its final destination has not yet been reached! However, the benefits enjoyed at Marconi Instruments in terms of reduced product cost, reduced development lead-time and increased sales revenue have certainly been worth the effort.

Process and Organizational Changes for the Implementation of Concurrent Engineering

Lucas Aerospace Actuation Division

Phillip Lewis

Introduction

A major review of the new product introduction process and the implementation of concurrent engineering at Lucas Aerospace Actuation Division produced significant benefits within the first five months, when a thrust reverser engine actuation system was developed using the new approach. This demonstrator project resulted in clear product cost savings of around 30 per cent.

Many new processes and working practices were adopted. These included the use of dedicated project teams, integrated computer aided engineering and quality function deployment. These new processes show how simultaneous working and close cooperation between design and manufacturing engineers can achieve significant reductions in cost, lead-times and quality problems. It also led to the successful development and implementation of a new project management structure in design and manufacturing engineering. The management structure was modified on the basis of maintaining a critical mass of key skills, reducing levels in the hierarchy, reducing wasteful activities, encouraging parallel activity and enhancing product focus.

Company background

The Actuation Division, based at Fordhouses in Wolverhampton, is a leading manufacturer of engine and airframe actuation systems such as thrust reversers and flap and slat systems. In 1989 nearly 1300 people worked at the Fordhouses site, including about 250 design engineers and manufacturing engineers. The decision to review the product introduction process was taken in April 1989, as part of a major programme of change throughout the company. The decision was driven by changes in market requirements. The company was moving from the supply of mainly military equipment to the supply of equipment for the civil aerospace market (mainly Airbus). In particular, this required shorter product introduction lead-times and very competitive pricing (as opposed to previous cost-plus contracts). A special Business Development Group was set up to direct the change process, consisting of five senior managers from the Actuation Division and two external consultants. The Business Development Group began with an initial target to reduce the cost of operating the business by 30 per cent.

The strategy agreed by the Business Development Group centred on the three key processes – manufacturing, commercial and new product introduction. A separate task-force (task-forces in the case of manufacturing) was set up for each process, with the company's overall business objectives broken down into clearly defined individual targets for each task-force. The manufacturing task-forces were charged with developing and detailing a product-focused manufacturing organization. This approach was based on the use of cellular manufacturing principles. The commercial task-force considered the areas of sales and product support (including repair and overhaul). The new product introduction task-force examined the process of designing, developing and improving product configurations and their impact on the manufacturing facility. Clearly, there was considerable overlap between the task-forces and this was tightly managed by a high-level steering group to ensure the effective and efficient use of resources.

By addressing each process as a single business area, the communication barriers between individual functions would be reduced, leading to less fragmentation, reduced waste and less duplication of activities. The Business Development Group also planned to introduce a new project and programme management system across the whole division, based on matrix management principles. This was to ensure that a flexible organizational structure was in place to meet the needs of customers. Matrix management allows this by enabling the allocation of resources to be reviewed continually and kept in balance with the ever changing needs of the business. The matrix management structure also ensured and facilitated the appropriate level of communication across processes. The matrix organization followed on from the work of the task-forces in each of the key processes and linked commercial, technical and manufacturing functions in a series of project teams controlled by senior programme managers reporting to the general manager.

New product introduction and concurrent engineering

One of the main starting points for the new product introduction project was the need to integrate the manufacturing and design engineering functions much more closely. The vehicle chosen to achieve this was concurrent engineering. This was reflected in the composition of the task-force set up to tackle the problem. The team included engineers from each of the major functions involved in product introduction. In particular, there were individuals from manufacturing, stress, design and CAE as well as an external consultant.

The task-force had a number of key objectives. These included reducing lead-times by 33 per cent, optimizing resources and reducing development phase failures by 50 per cent. It also aimed to achieve significant reductions in product cost by introducing design for manufacture and producing 'right-first-time' manufacturing instructions. The task-force aimed to raise the awareness of cost issues among the design engineers and to involve the manufacturing engineering function much earlier in the design process. The reduction of lead-times was also identified as an important objective. In addition to the obvious issues of competitive customer service, shorter lead-times would make an important contribution to overall costs. A more streamlined product introduction process makes much more cost-effective use of costly development resources, for example, and effectively extends the life and earning potential of the product.

The need for change in new product introduction

The project began with a detailed, four-week data collection exercise. The aim was to find out where resources were being used and where quality and response could be improved. An initial, qualitative analysis of the existing systems revealed a number of problem areas. A great deal of effort was wasted on duplicating activities across different functions, and the cumbersome procedures needed for tasks such as engineering changes were seriously slowing down the process. Design engineers had insufficient access to valid cost information, and manufacturing engineers were excluded from design until much too late in the process – after the major design commitments had already been made.

A detailed analysis of design engineering and planning tasks confirmed these conclusions. Less than 30 per cent of total design effort was being spent on new products, most of the remainder being expended on amending existing designs. Over 70 per cent of all documentation issued related to product modifications rather than new products. This was primarily as a result of poor communication between design engineering and manufacturing engineering. For example, it would not be unusual for design engineering to specify tolerances that could not be 'held' in the manufacturing facility. To illustrate the scale of the problem, it was found that on average each drawing was modified about 2.5 times.

By simplifying change procedures, administrative overheads could be considerably reduced. For example, by reducing (or eliminating) the level of documentation required to initiate change in the concept design phase, early change would be encouraged and administration reduced. Even more importantly, by involving manufacturing at an earlier stage of design, many of the modifications could be eliminated altogether. This would reduce minor modifications by around 40 per cent. In turn, this would cut the total design engineering effort by around 30 per cent. A similar situation prevailed in the manufacturing engineering department. More than 30 per cent of all planning effort was spent on modifications, a large proportion of which could be eliminated by more up-front involvement in design. The team also identified a high level of unnecessary clerical tasks, amounting to as much as 20 per cent of the total manufacturing planning effort. Figure 3.5 summarizes many of these key points.

Alongside the task analysis, the team also studied the performance of the current product introduction process, measuring lead-times, adherence to schedules and overall costs.

Design Engineering
- A simplified change procedure for amendments will result in reduced administration time (40% of total effort).
- Up-front involvement of manufacturing has the potential of reducing modifications by approximately 40%.
- Reducing modifications by 40% will reduce design engineering effort required by 30%.

Manufacturing Engineering
- Approximately 30% of all process planning effort is spent on modifications.
- 20% of effort is clerical and the tasks could be performed on the shopfloor.
- At least 10% of effort could be saved by up-front involvement.
- Design and manufacturing engineering both spend approximately 70% of total effort on projects that are already in manufacture.

Figure 3.5   **Task analysis conclusions**

The team also examined the quality of design, expressed in terms of rework levels, modifications and concessions. From the study of several projects, a number of consistent trends began to emerge. Long-winded checking procedures, with many levels of authorization and the resulting block releases of drawings (checking took place in batches) were clearly holding up the flow of work through the system, and the process was generating constant modifications. Lead-times were also being extended by a high level of changes at the detailed design and process planning stages, with process planning layouts needed for around 40 per cent of all detailed designs.

The data collection exercise clearly pointed to the need for the implementation of concurrent engineering. Concurrent engineering would address the three main priorities which had emerged. First, products were not being designed to suit the manufacturing process. There had to be more up-front cooperation between design and manufacturing engineers. Secondly, there was a need to produce an integrated set of manufacturing instructions for each product. There was a necessity to move away from the concept of detailed drawings and planning layouts as separate documents. This would open the way towards genuine concurrent working in manufacturing and design, and would provide more efficient support for the manufacturing facility. The third major priority was the existing functional structure of the company, which was creating major bottlenecks. The current checking procedures were obstructing the flow of work between departments, and engineers were concentrating on departmental rather than project targets. There was a clear need to move towards a project-oriented, team-based structure.

Specifying concurrent engineering

The redefinition of the new product introduction process was the key first step. The team began by drawing up a detailed specification for the process. The design process would become an integrated, multidisciplinary operation, with direct inputs from design, manufacturing, service engineering and suppliers.

Wherever possible, departments should work concurrently on the different phases of design and manufacturing engineering, and changes should be concentrated in the earliest possible phase.

The team also recognized the need for tighter planning and control, estimating resources in detail at the outset of a project, setting realistic targets and continually monitoring progress with interim targets. This approach would identify potential cost and schedule overruns before problems got out of hand. The key elements of the specification included the following requirements:

- to ensure meaningful up-front involvement of all necessary disciplines in the design process;
- to undertake parallel working to reduce lead-times and the need for downstream changes;
- to introduce formal procedures that actively encouraged design iteration in the early stages;
- to ensure levels of communication that actively enhanced project team integration;
- to eliminate duplicate tasks;
- to stimulate cost-effective designs by the continuous feedback of actual costing data;
- to provide good planning, monitoring and control by:
  - enabling early and accurate resource requirement estimates
  - providing interim monitoring milestones
  - rationalizing sign-off procedures;
- to provide the right data at the right time in the right format for the receiver.

In addition, the necessity was recognized of introducing best practice principles and methods, such as finite element analysis and solid modelling. Each technique and tool was evaluated for its relevance and integrated into the appropriate phase of the revised new product introduction process. Techniques were not just specified on the basis that they were the latest 'must have' management fad. The basic specification led to a revised design flow, replacing the existing complex sequence of processes and checking procedures with a simple, parallel flow of operations through design and manufacturing engineering – see Figure 3.6.

In place of the conventional division into design and manufacturing engineering, the revised process was split into two major phases – the design phase and the product data package (PDP) phase. Within each phase, design, manufacturing engineering and quality functions work together towards the same milestones – producing design, manufacturing and qualification support documents in the design phase, and creating a complete set of manufacturing instructions and tool designs as part of the product data package. Detailed concept and prototype testing run concurrently with the main development phases.

This emphasis on parallel working was an important part of the move towards a project team structure. Engineers should no longer be working exclusively to meet functional objectives within their own department; they share a common goal with the other specialist functions within the project team.

The design phase

The design phase involves the production of detailed support documents for design, manufacture and qualification. In the design support document the new product introduction project team specifies processes, materials, tolerances, standard parts and design layouts. The design support document consists of a drawing defining the shape of items in sufficient detail to enable piece part manufacturability to be assessed, project stress analysis and weight estimation to be undertaken, performance requirements to be confirmed and cost estimation completed. This drawing is termed the 'bodyline' design. A parts list indicating piece part attributes such as material selection, process selection and supplementary geometrical information such as gear data is also compiled. The first pass 'bodyline' design is then developed through the interaction of the various team disciplines. The manufacturing support document is produced concurrently, taking inputs from design and manufacturing to address key issues such as process planning routes, the make in or buy out decision, the use of manufacturing facilities and tooling.

79

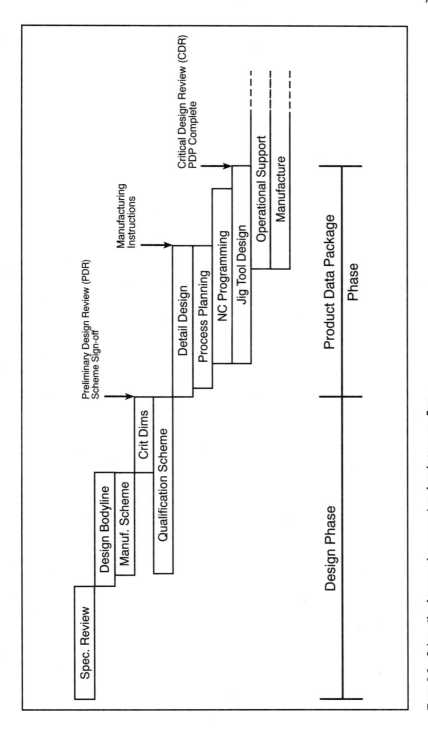

Figure 3.6 **Schematic of proposed concurrent engineering process flow**

A design-to-cost iteration process takes place with the team endeavouring to optimize the 'bodyline' configuration. The team manager will monitor cost performance against the target cost set by programme management. In parallel to the manufacturing evaluation, other key disciplines will be affecting the 'bodyline' configuration. These include:

Stress and performance	Service engineering
Systems	Supplier
Reliability and maintainability	Materials
Development	

Daily workshops are convened and represented by the principal disciplines, namely conceptual and detail design, manufacturing and stress. Other team members are considered to be on call unless they have identified a problem or idea that needs to be discussed, in which case they highlight the requirement to the principal engineers.

The live configuration is made clearly visible on a view board geographically central to the team, along with the parts list and a clear indication of the current issue status. Master design 'bodyline' drawings also reside here, enabling easy access by all team members for reprographic purposes. This means of communication will allow people outside the main team activity to keep in touch with the evolving design. A team meeting is called each time the issue status of design 'bodyline' drawings is raised, or at least once per week.

It is essential that suppliers are involved at this early stage. They must be recognized as representing valuable expertise and similar workshops involving them are conducted to minimize the risk associated with their product. To ensure their participation, the suppliers are required to generate source control drawings for their parts. By ensuring that suppliers have to produce design drawings, the team can be sure that suppliers have thought through issues in the same depth as internal engineers. At this stage they will be preliminary, awaiting critique and approval.

In parallel with this 'bodyline' optimization activity, the qualification support document is initiated. This is designed to ensure the generation of sufficient data up front to support the compilation of the detailed test rig specifications and test documentation during the product data package phase. The qualification support document contains the following sections:

- a 'specification response' indicating what test rigs are required to undertake each element of the qualification process;
- the specification of the number of test units required;
- a 'qualification test matrix' indicating the range of tests to be undertaken;
- a 'qualification test programme' indicating the timescales associated with the tests;
- the qualification procedures that will be used to ensure adequate testing and recording of results.

Once the 'bodyline' has become more defined, the team manager will call a preliminary design review (PDR). The purpose of the PDR is to get the whole project team to 'sign off' the schemes that have been generated. Traditionally, PDRs were dominated by design engineers with a token manufacturing engineering presence. There might well have been a dozen design engineers and one manufacturing engineer in attendance. The design engineers would have intimate knowledge of the proposed design; the manufacturing engineer would be coming to the review 'cold' with little knowledge of the design and the principles behind it. It is not hard to see why a meaningful manufacturing input would not be made. It would take a very able manufacturing engineer to make a useful input on such limited exposure to the design, and a very brave manufacturing engineer who would be prepared to withhold his sign-off in these circumstances. The new PDR was designed to ensure meaningful manufacturing input and ensure the problems associated with change did not occur needlessly.

With manufacturing methods and component sources identified from an early stage, the design phase also produces a detailed and accurate costing scheme. This includes estimates for both the development project and the final product. By the end of the design phase, the programme manager is equipped with a detailed analysis of risk and project costs, helping the management team to maintain full control over schedules and budgetary targets.

The product data package phase

The design and manufacturing support documents are then used as the basis for the product data package (PDP). This phase brings design and manufacturing engineering together, to create a detailed set of manufacturing instructions. The instructions provide all the information needed to manufacture the finished product, including process layouts, detailed tooling and component dimensions, NC programs and assembly instructions. With all the relevant data produced as part of the same, integrated process, this system eliminates constant reiterations of drawings and planning layouts. Alongside the manufacturing instructions, the PDP phase generates a full technical data package, involving service engineers to produce detailed maintenance procedures and manuals. This phase terminates in a critical design review (CDR) to sign off the complete data package.

Implementing concurrent engineering

Having specified a concurrent engineering process, a top priority was to find out how well the outline flow would work in practice. An eight-week demonstration project was undertaken. The product selected for the demonstration was a thrust reverser engine actuation system. Under the leadership of a project manager, a full project team was brought together, assembling engineers from each of the functional departments.

The results were significant. Compared to the original bid estimates, the new concurrent engineering process led to direct component savings of more than 30 per cent, together with a reduction of nearly £4000 in tooling costs. Projected forwards, these figures lead to a 15 per cent product cost reduction across the whole project, with savings of around £320 000 over five years. Example changes and savings included material specification changes resulting in the need for processes such as electroless nickel to be eliminated, and the specification of existing proprietary components rather than the design of new specialized components.

The experience of the demonstrator implementation also indicated that the new flow could significantly cut the number of engineering changes. On the demonstrator implementation, there were only around half the changes expected on a conventional project. If this figure were repeated, the new concurrent engineering process could reduce the total design effort by as much as 30 per cent on future products.

As soon as the demonstrator results were assessed, the company began to make plans for changing the whole product design operation over to the new concurrent engineering process. A second trial project (an airframe flap and slat actuation system) was started immediately, and a timetable was set for the changeover of every project within 12 months.

Changing the organization structure for concurrent engineering

The next major stage of the project was the development of a new organizational structure. There was a deliberate policy to wait until the new concurrent engineering process was implemented before specifying the organizational structure. It was felt to be important that the structure should be derived directly from the process, with the flexibility to adapt to the needs of different projects. Although a matrix organization was needed, it was realized that the project team structure would not be the right approach for every product.

The process began by identifying a series of basic criteria for the new organization. In line with the overall objectives of the change programme, a matrix structure had to be established, with individual project teams formed across functional boundaries. The structure had to be designed around the new process flow, streamlining the preparation and release of new documents such as design support documents and manufacturing instructions.

The existing management structure, shown in Figure 3.7, was seen as too complex and unwieldy and the new organization would have to reduce the number of management levels, and encourage tighter control with shorter lines of reporting. At the

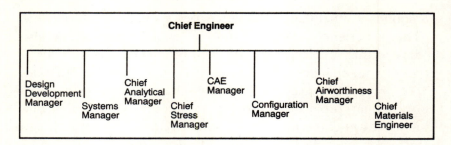

Figure 3.7   **Original organization structure (top level)**

same time, there was a clear need to maintain opportunities for career development. A 'balanced view' was also an important factor, with equal 'power' given to each of the major disciplines. The company needed to move away from the traditional design engineering-dominated approach to product introduction, with equal weight given to both manufacturing and design engineering. The structure also had to encourage flexibility, with engineers given plenty of opportunity to develop multidisciplinary skills. Unlike the old structure, where departmental targets had often conflicted with project goals, performance measurements had to encourage cooperation between functions.

The starting point for the new structure was the formation of natural groups, at both the functional and the project team level (the two axes of the matrix). The new concurrent engineering process provided the basis for the new organizational structure. The restructuring operation began with a review of the functional departments and the existing management structure. By reducing the number of departments from nine to six, the new structure created a more natural division of functions. The milestones identified as part of the new concurrent engineering process flow were used as the basis for the new functional groups, with engineers within each department working towards the same project targets. By combining design and manufacturing engineering, for example, the whole process of preparing manufacturing instructions was brought into the same functional department. Major project tasks could now be handled within the same department, without the cumbersome procedures needed to transfer data between functions under the old structure. At the same time, the management levels were rationalized to provide a more efficient and logical chain of responsibility. The new management structure is shown in Figure 3.8.

Alongside the new functions, a structure was set up for seconding engineers to individual project teams. Each project team draws engineers from each of the functional departments (headed by a functional manager). Each project team is led by a project team leader. In turn, the team leader reports to a programme manager. The programme manager is responsible for setting milestones, schedules and budgets, and allocating resources. The project team leader controls the technical integration of the project, and reports back to the programme manager with detailed information on milestone progress and resources. Figure 3.9 shows the matrix structure that was adopted. The quality and technical control of each functional task is the responsibility of the functional manager. The functional managers will also oversee the overall development of skill resources through training and recruitment, and assist the programme manager in estimating costs and resources. Appraisals are undertaken by functional managers with a significant and determining input from the project team leaders.

The new structure has been fully implemented across the company. It has given the Actuation Division a simpler and more efficient approach to product introduction, with non-value-added activities reduced to a minimum. It has also opened the way to the implementation of new technology and methods such as computer aided engineering.

Conclusions

The company undertook some very detailed analysis to understand its current situation. A concurrent engineering solution that fitted the needs of the company was then developed. The implementation of a panacea, 'must do' approach to concurrent engineering was consciously avoided. The concurrent engineering process adopted was developed as a consequence of:

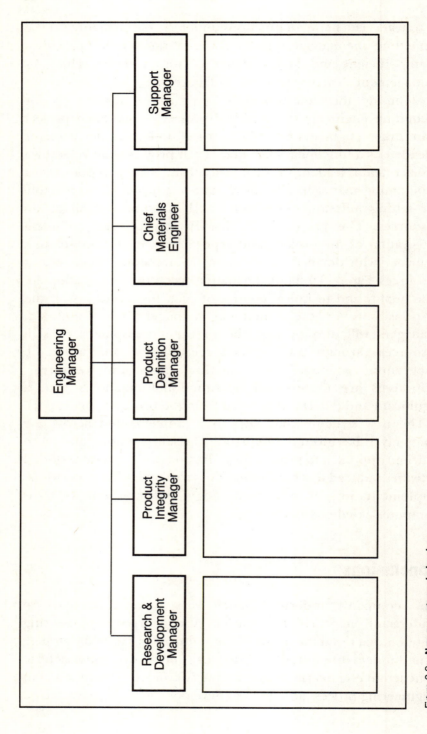

Figure 3.8 **New management structure**

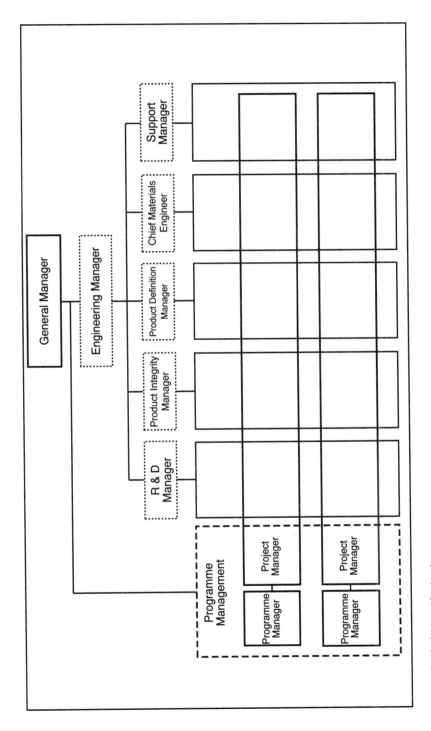

Figure 3.9 **Project matrix structure**

- customer requirements
- product technology and complexity
- the skills of existing personnel
- organizational constraints.

A pragmatic approach led to the introduction of best practice, bringing significant benefits to the new product introduction process. Lucas Aerospace has concentrated on a step-by-step approach to restructuring and simplification, focusing on concurrent engineering. The results have given the company a sound, long-term foundation for competitive performance.

4
The Information Technology Approach

Instron Ltd, Design to Distribution Ltd (D2D) and IBM UK, Havant

In contrast to the previous chapter which emphasized structural and process change related to new product introduction, this chapter concentrates on the experience of three companies which have invested heavily in information technology (IT) support tools. The companies, Instron Ltd, Design to Distribution Ltd (D2D) and IBM UK, Havant, have all developed sophisticated electronic communications capabilities in order to facilitate rapid evaluation of product designs by development team members who are often located at geographically dispersed sites. Rather than physically locating the team in one place, the application of information technology has allowed these companies to develop 'virtual teams' which communicate electronically. The actual physical location of each team member is less of an issue than the ability to access common databases and product development software.

In relation to the concurrent engineering framework (see Figure 4.1), the three companies can be seen to be operating under the influence of slightly different combinations of pressures. All of them are naturally experiencing pressure for Efficiency, but in each instance this is combined with Incremental Change and/or Proficiency. The particular case study which demonstrates a significant pressure for Proficiency is that of D2D. This electronics manufacturing services provider must be able to demonstrate its capabilities to its customers before it will be awarded a contract. There is therefore a strong requirement that the company can demonstrate a high level of proficiency in its operations.

Instron and IBM Havant must also react to the pressure of Proficiency by retaining and enhancing their skills and expertise in order to maintain capabilities in product development. However, both companies additionally see a strong pressure for Incremental Change in their products. For IBM Havant and Instron, the pressure to develop new products on a reduced timescale predominantly relates to incremental improvements to earlier products.

The pressure for change

The three companies discussed in this chapter can be considered as reacting to long-term trends in product capabilities and life-cycles. The functionality of future products, development timescales and the time before updates are required are all fairly predictable. This is because all of the products described in the

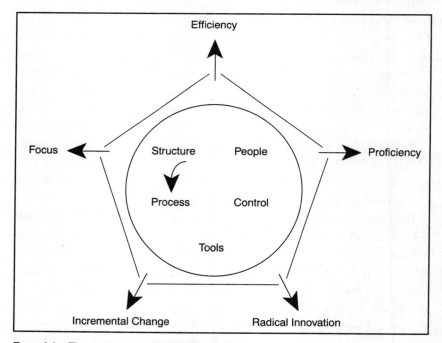

Figure 4.1 **The concurrent engineering framework**

case studies are based around microelectronics technology. The future capabilities of microprocessors are seen as following historical trends in terms of processor speeds, cost and development time. As a consequence, the future functionality of products already based on this technology has a high degree of inherent predictability. In this respect, therefore, the case studies in this chapter describe companies which are currently following evolutionary paths to improvement where the pressure is for Incremental Change and not for Radical Innovation.

The change

IBM Havant describes the requirements for an IT infrastructure with particular reference to the development and defect prediction of computer workstation subassemblies. The significant reduction in development time through statistical analysis of test data dramatically shortens the overall time to market. For D2D the primary pressure is to react to the requirement to demonstrate its development capabilities to potential clients. It must be able to show proficiency in terms of new product development to instil a sufficient level of confidence in the client in order to obtain the contract.

There are significant differences in emphasis in terms of the areas of IT that each of the companies focuses on in its case study. IBM Havant has concentrated on developing a common database infrastructure accessible by all departments. D2D has a similar requirement for engineering data management, but its emphasis lies more on providing an electronics manufacturing service to clients and its IT strategy reflects this fact. Instron, on the other hand, has responded to the need for communication between its two major sites, one in the UK and the other in America, in order that full cooperation is ensured for common development projects.

For all three companies the striking feature is the lack of recent dramatic change in their concurrent engineering implementation. Unlike those situations which require change to organizational structure and processes, these three companies are following evolutionary paths to achieving their goals in new product introduction. To a large extent this reflects the fact that the more dramatic changes have been carried out in the past and the companies have now moved on to the new priorities associated with IT. However, it does also serve to demonstrate that once the IT route has been adopted there is a reduction in the requirement to consider organizational issues related to personnel location, with the focus falling on electronic infrastructure development.

All three case studies illustrate the fact that the companies involved have a very clear view of the process of developing new products. Formalized processes are specified which are followed in all new product developments. Where a service supplier such as D2D is concerned, then the procedures and processes are available to be slotted into the client's requirements. For Instron and IBM, which are effectively developing their own products, the procedures can be broken down into very clear stages which every project will follow from conception to launch.

To a certain extent the issue of 'virtual teams' is hidden in the case studies since it has become such an accepted fact of life within the companies. This can be seen especially in the IBM case study, which describes the matrix organizational structure established to develop new products, but goes on to say that 'it is common for staff to have more contact with their project team than with the person seated next to them'.

In all three case studies the creation of multidisciplinary teams involves bringing together personnel from various skill areas. The organization of the business therefore retains functional centres of excellence from which the teams take their required resource. The importance of retaining these skill areas in order to ensure long-term skills retention is emphasized particularly in the Instron case study. A previous attempt to reorganize exclusively along project lines was reversed to retain strong centres of technical knowledge.

Finally, the importance of rapid and reliable communication between the various team members involved in product development cannot be overstated. In all the case studies the emphasis on electronic data management has encompassed the establishment of electronic links to ensure rates of interaction which support a team-based approach to product development.

An Evolving Product Introduction Process

Instron Ltd

John McAllister and Chris Backhouse

Summary

The introduction of concurrent engineering within Instron Ltd is seen as a natural continuation of evolutionary change that has being progressing for many years. Modifications to the process of introducing new products have generally been the natural response to a changing set of market requirements. Reliance on a range of standard products has been modified to accommodate the market's need for customized materials testing systems. The company operates parallel teams, in the UK and America, for product development and has done so for many years. The coordination of the two teams is based on a combination of dedicated electronic communication links, the use of documented standard procedures which comprehensively define the process to follow in developing a product, and product module interface specifications. The company operates with a flexible matrix management structure while retaining functional lines of responsibility.

Background

Instron's origins date back to the early 1940s with the involvement of the two founders in testing material for parachute harnesses at the Massachusetts Institute of Technology (MIT). It was recognized that special equipment was required to provide constant straining rates despite increasing loads. No equipment with the necessary capabilities was commercially available at the time and a purpose-built machine was designed to conduct the experiments. Subsequently a company was created to design and manufacture test machines. From this early start, Instron Corporation developed a strong lead in the development of materials testing equipment, a lead which has remained with the company to this day.

Instron Corporation grew steadily in size as the materials testing market expanded, until by 1960 the number of employees exceeded 400. A decision was then taken to extend into the European market by establishing a marketing, design and manufacturing facility, Instron Ltd, in the UK. The size of the UK operation rose rapidly to match that of the American one by the end of the 1970s. A Japanese operation was initiated in 1965 and currently employs 50 personnel involved in sales, service and customizing for the local market. In the 1990s a broadening of the corporation's mission saw the acquisition of companies in the fields of environmental and hardness testing. In total Instron now employs approximately 1200 people and has a turnover approaching £100 million.

For many years, until the late 1980s, there was a steady growth in the size of the market for standard test equipment. More recently, however, this market has reached maturity. The result has been that competition has become price dominated, especially as a consequence of the recent recession which, for test equipment, was global in nature. Customers, having once seen prices fall in real terms, are unwilling to accept subsequent price rises as the economic environment has improved. As a consequence the company has expanded into other, related areas offering higher value-added potential.

Instron has extended its activities to offer complete laboratory test services. This has been achieved through a variety of initiatives. Acquisition of companies which manufacture related test equipment, such as hardness testing, has enabled Instron to satisfy customer needs for the complete range of equipment likely to be required in a single materials laboratory. The customer then benefits from a single source supplier who can provide maintenance support for the complete laboratory. In addition, Instron has moved into the customized product field where special test rigs and sophisticated data management of test results are required. This is a growing market as companies move along the road to meet the requirement for total life-cycle tests rather than simple strength tests. Finally, Instron has moved to update product designs constantly and introduce new standard products. Where this has focused on developing new controllers, it has had the additional benefit of providing the facility to retrofit onto its competitors' machines in addition to its own. This is especially relevant to a large proportion of potential customers who already own a test machine, where any new product must provide considerably enhanced functionality at a price that the customer is willing to pay. There are significant international competitors, based in America, Germany and Japan, but strong technologically driven local competition is always present.

Technology

Instron's competitive edge from the outset has been excellence in the fields of control and sensor technology allied to strengths in mechanical and electronic engineering and materials science. Controllers, which are key differentiating products for Instron, have always employed leading edge technology, progressing from thermionic valves in the 1940s, through discrete then integrated circuits, to microprocessors and digital signal processing today. The latest controller, the 8500 Plus, was launched in 1994 and was the culmination of a product development process which encompassed parallel teams in the UK and America working on the design of both hardware and firmware.

Changes to the product introduction process

Until the early 1980s product development was carried on in a serial mode, with manufacturing often becoming closely involved in the process only when the engineering team had completed the design and documentation. This approach clearly could not continue as the market became more demanding and cost conscious. In response, Instron moved to ensure that all project teams were structured to include marketing and manufacturing input from their initial inception. This requirement to develop cross-functional teams at the very earliest stages of a product development programme was written into the formal company procedures, as can be seen in Figure 4.2. As soon as the initial feasibility study was initiated to develop a project budget, a cross-functional team would have to be considered.

The input of manufacturing personnel to the design process and the generally higher requirements being placed on them resulted in a change in the required educational profile. Instead of simply being involved in planning the manufacture of a new product from given designs, they were required to contribute actively to the design from the earliest stages. Required qualifications of manufacturing engineers were therefore raised to degree level, whereas previously the company would have been satisfied by an engineering diploma level of education. Manufacturing engineering personnel were relocated into the same office suite as that occupied by the design engineers, although, as is described later, they remained in their functional groupings.

Whilst there were benefits in closer integration of design and manufacturing in terms of new product development there were, and remain, inevitable conflicts in priorities. In particular this relates to resource allocation of manufacturing personnel between the activities of new product development and those of producing existing products. In the early 1980s manufacturing activities had actually been divided to support operations and development as separate activities. However, this had led to a perception of two tiers of roles and crucially did not facilitate feedback of mistakes. The decision to split the activity has

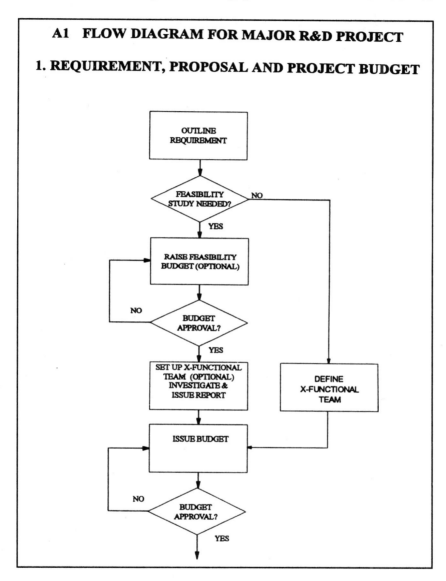

Figure 4.2 **The documented procedure to develop a project budget**

therefore been reversed, and the conflict of resource allocation has been accepted as the 'lesser of two evils' to be overcome through consensus management.

One of the major factors that facilitated the introduction of concurrent engineering techniques was a change in culture that the company had been experiencing over the previous ten years. As the market began to demand more customizing of products, then the roles of individuals started to blur. Informal teams set up to satisfy a particular contract slowly emerged as the standard way of progressing such projects. In practice the changes that took place were so slow that it would be difficult to identify a particular time when they were initiated. When such change is imperceptible then there is not likely to be resistance, and unlike many companies moving over from a functional to a more project-oriented structure, Instron followed a smoother path of change. The term concurrent engineering is therefore seen simply as a suitable description for the approach to developing new products that Instron had naturally been adopting. The process of change has now been accepted by most as a way of organizational life within the company, as new project groups are formed and others disbanded to meet contract requirements.

Design tools

A key event in the adoption of concurrent engineering techniques within Instron was the introduction of CAD. The corporate decision to invest heavily in an advanced system *common to both sites*, motivated by efficiency gains, was crucial. 3D modelling, initially wire-frame, was embraced from the outset. 2D drawings became the output from the design process rather than the means by which it was conducted. For the first time sales, marketing and manufacturing could visualize the end product at the feasibility stage and know that they were all talking about the same product. Indeed, sales realized that the powerful tool which engineering had for design also provided them with a powerful tool for selling.

The move in 1991 to a CAE system offering analytical tools such as finite element stress analysis, modal analysis and kinematics within a true solid modelling environment served to unite analysts and designers. It was also a vital step in reducing project design and development time while increasing the ability to 'get it right first time'.

The past few years have also seen a significant increase in the use of electronic communication tools within Instron. Full use is now made of electronic mail both within and between Instron in the UK and America. A direct high speed data link between the two sites was installed 18 months ago and now allows the rapid transfer of geometric and text data to facilitate rapid turnround in design concepts. In addition, both sites are capable of accessing controlled databases of product modules which specify supplier information, change levels, bills of material etc. Other sources of information required by the designers have been automated and can be accessed across the network. Databases of project files, which are the repository of all of the design information for a project, standards and technical reports are available. The network also provides statistical feedback on service reports and holds warranty information against model type.

Whilst the information network is not seen as the primary route to operating in a concurrent engineering mode, it does provide the background environment to facilitate this approach. High speed data links, by which designs can be considered and modified very rapidly, ensure that co-location of all team members is not necessary. Working from common databases which contain controlled information ensures that all personnel are operating to the procedures and utilizing the same standard parts.

Company structure

The structure of both the UK and American sites are basically functional in form with additional matrix-type reporting lines. The overall structure of the UK site is shown in Figure 4.3. The managing director of the UK site reports to the American corporate headquarters. UK directors report to their managing director but also to corporate headquarters' functional directors. Beneath the UK site director level there exists a similar functional divide where personnel are located within traditional departments such as engineering, manufacturing and marketing.

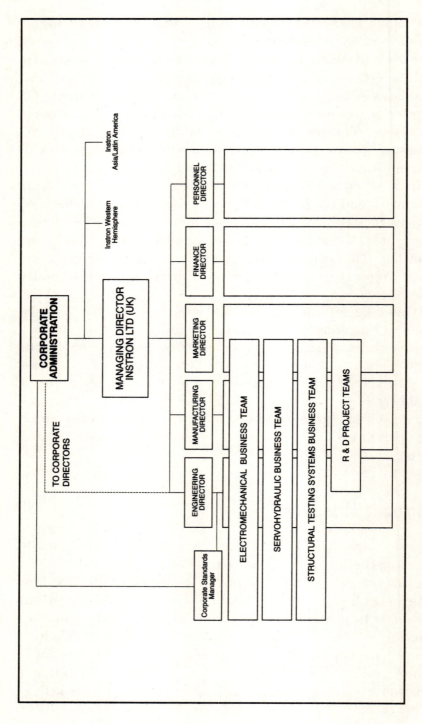

Figure 4.3 **Project groups overlay the company's functional structure**

The functional managers are responsible for controlling the reward system for all employees within their grouping and all the reporting system at this level follows the functional structure. In addition there are corporate employees who may be sited either in the UK or America. These include a corporate standards manager and corporate product planning (more of these later).

Overlaying the functional structure is a flexible project-based structure. Interdepartmental business teams exist in the areas of electromechanical machines, servohydraulic machines and structural testing systems. Their role is primarily concerned with marketing and sales and with the design and development of custom products (not with standard products). The business units are created by co-locating project and design engineers, manufacturing engineers and marketing personnel.

An interesting aspect of the business team structure is that the individual members retain their normal departmental reporting lines. Since all members of the business team originate from defined functional groupings, their respective responsibilities are obvious and individuals take on actions as appropriate. A company director acts as 'mentor' to each business team but not in a directing capacity. The mentor provides advice and a direct communication link to top management. The mentor also acts as a facilitator on resourcing issues both internally and externally, e.g. strategic alliances.

R&D project teams are the groups of employees responsible for new standard product development. This may be an existing product which could be reduced in cost through redesign, or the development of a completely new product. The membership of the project teams comprises personnel from engineering, manufacturing, sales, service and occasionally finance, and is therefore somewhat similar to that of a business team. However, the significant difference is that all the project team members are not necessarily co-located. Whilst design and manufacturing engineers occupy the same office areas, they remain in their functional groupings. Other team members such as marketing and purchasing personnel are located at other points on the company site. In addition, where the UK and American sites are working on the same project, as was the case in developing the

8500 Plus series controller, then co-location is clearly not a feasible option.

Another argument against co-location is the considerable cross-activity between the team members. Individuals will often contribute to several project teams. Since projects range from single component redesigns to the development of completely new products, the degree of input required from each team member will vary considerably between projects. Whilst the major players in a given project team will interact on a daily basis within the functional groupings, they may also interact on a less regular basis with a team for which their input is less time consuming.

Leadership of the project teams is provided by product group managers who report directly to the engineering director. External monitoring of the project is achieved through formal review meetings corresponding to the project milestones as documented in company procedures. An example of this is shown in Figure 4.4 for the development activities up to the engineering prototype, followed by units for formal evaluation by marketing and manufacturing. It should, however, be remembered that both manufacturing and marketing personnel have had a significant input into the design in their role within the cross-functional teams.

Design reviews are seen as formal acceptance milestones. Additionally, in order to ensure complete agreement and consideration of all detail a project review may be held twice, once on the UK site and once on the American one. Major project team members participate in both reviews.

'Voice of the customer'

At the time when the initial concepts were being developed for the 8500 Plus series, a marketing initiative commenced from within corporate product planning. This section of Instron is responsible for identifying long-term product ideas and for convincing the business as a whole on future product strategies. Whilst it was felt that Instron was satisfying current customer requirements, there seemed to be an opportunity to identify

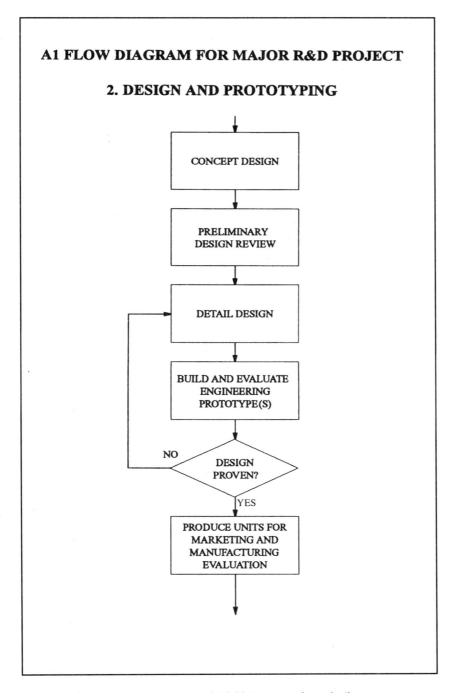

Figure 4.4 **Formal procedures for developing initial prototype for evaluation**

other potential requirements, the benefits of which the customers themselves had not yet identified. To these ends a programme of activity under the banner of 'voice of the customer' was initiated.

An important concept within the 'voice of the customer' initiative was that employees from throughout the business, and not just from corporate product planning, conducted the visits. It was seen as fundamental to gaining a close understanding of customer requirements and to ensuring that ownership of the final results would be spread throughout Instron. The knowledge and understanding generated within the engineering functions was directly responsible for ensuring that common purpose was maintained in all subsequent development projects. The voice of the customer was therefore viewed as a fundamental requirement for the introduction of improved performance within the area of new product introduction and for concurrent engineering to operate successfully.

The voice of the customer consisted of personal visits to an array of existing users of test equipment. Two Instron employees, typically one from marketing and one from engineering, would visit a user to determine their views. Interviews were based on a structured questionnaire, although the individual questions were intentionally non-specific in nature. They asked questions which sought to identify what the user would like from test equipment rather than what the customer felt was lacking from current products. Through a large number of such interviews a corporate document was created which detailed the likely long-term development requirements of new products in terms of customer requirements.

The first benefit from the study was the identification of a specific market area that Instron could immediately attack. One of the most intractable problems found to be experienced by customers concerned situations where they operated a range of machine controllers and application software having differing interfaces and requiring extensive training to master. Customers expressed a strong preference for consistent controller and software interfaces within their test laboratory, even if the test machine structures themselves came from different suppliers. Instron now have a significant market in retrofitting competitors' equipment with new digital controllers and application software.

The second benefit was the documented expression of a large number of customers' requirements. As with any technology-based company, many of Instron's employees had their own ideas about what constituted important product features. With the customer requirements better understood and documented, debate and uncertainty in this area was greatly reduced. This knowledge of customer requirements now drives new product development and is actively kept up to date.

Corporate standards

The responsibility of the corporate standards manager is to maintain and ensure the adoption of all company product standards. These standards define in comprehensive detail all of the interface requirements between components and subcomponents of mechanical equipment, controller hardware and software products. They define physical sizes, connection types, communication protocols, voltages etc. Thus precise specifications of all possibly occurring interfaces are defined, such as those between actuators and load cells, between controllers and test machines and between transducers and control electronics. Unlike many companies where standards do exist these are followed precisely. No product is developed within the company unless it complies completely with the standards. In addition, the corporate standards manager ensures that the documentation specifying all bought-in components for all products is maintained and is up to date. Components cannot be purchased unless they exist on the corporate database which can be accessed from all Instron sites. Therefore common parts will be employed in products irrespective of the site at which they are designed and manufactured.

Conclusions

Instron recognizes that the functional division of the company is not the conventional picture of a concurrent engineering implementation that many people would expect. However, the potential problems associated with functional divisions are well recognized and the procedures and methods of working that have been established within the company have been designed to minimize any such problems. In addition, the company remains a technology-oriented organization. Whilst the skills inherent in developing testing equipment and systems can be learnt, they are extremely difficult to master. The knowledge necessary to develop test equipment is built on 50 years of experience. The primary core competence of Instron is therefore the ability of its personnel in marketing and in design and development. For this reason it was decided that Instron should protect and develop knowledge through a primarily functional-based structure.

In light of this decision to retain functional structures, the introduction of concurrent engineering can be seen to be procedural- and information technology-dominated. Fully documented procedures for developing new products ensure that each project follows precisely the same basic path. As part of this path the establishment of cross-functional teams is considered at the earliest stage. All subsequent activity is then carried out by the cross-functional team to ensure that significant marketing and manufacturing input occurs during the early stages. Project reviews occur at regular intervals to provide a project monitoring system.

In support of the documented procedures, product databases are maintained to ensure that all designs conform to standard company specifications for parts and for module interfaces. As a consequence geographically distant project teams can work in parallel, being confident that their part of the design, conforming as it does to interface standards, will be compatible with other components being developed concurrently.

Finally, the implementation of a comprehensive approach to information technology provides the linkage between people that is traditionally achieved through co-location. In particular, high speed data links which make feasible the transmission of comprehensive design detail enable interaction to take place at a rate exceeding that obtained through traditional meeting schedules. Communication across the Atlantic Ocean is little different from that between various locations on the same site.

The future

Whilst the previous description of the product introduction process has demonstrated why a functional organization continues to work for Instron, it is not the whole story. Clearly, whilst the adherence to company standards provides a significant control mechanism to ensure right-first-time design, it also implies certain cost penalties. Specification of interface designs for all circumstances indicates redundancy and consequently extra cost. In the past the benefits of ensuring a smooth development programme have outweighed the extra cost of interface standards. One of the lessons to be learnt from other industry sectors is that it is unlikely that this situation will last forever. It is therefore assured that Instron will continue to evolve its organization, and the way it develops new products, for some considerable time to come.

Electronics Manufacturing Services: The Information Technology Infrastructure

Design to Distribution Ltd (D2D)

Mike Meadowcroft and Naomi Brookes

Summary

D2D provides electronics manufacturing services to client customers. As such it provides that vital concurrent engineering element: up-front input of manufacturing information into the product development process. To achieve its present and successful position in a highly competitive market D2D has invested heavily in its information technology infrastructure. It has developed a comprehensive IT capability allowing for design information to be considered by all team members, however geographically dispersed they may be. Supported by corporate component databases and automated design validation packages, the company can ensure that designs move rapidly from concept to manufacture with very high levels of confidence through its application of engineering data management (EDM). The company's product development infrastructure is now so developed that the organization is constructed around virtual teams, with physical co-location being relatively unimportant.

Introduction

Design to Distribution Ltd (D2D) is a wholly owned subsidiary of ICL. It was created on 1 January 1994 when the original Manufacture and Supply Division of ICL was established as an independent company. It currently employs approximately 2500 people on five sites within the UK and has a turnover of £300 million. D2D's business is primarily in manufacturing electronics products, but as its full name suggests it provides services which range across the spectrum from the design of products to their distribution to customers. In fact, it extends beyond this range to encompass disposal and recycling.

Although it only obtained true independent status in 1994, the company had been operating as such for some years previously. This is reflected in the gradual change in the proportion of its activities related to ICL business. In 1990 all of its business was involved in the manufacture of ICL products. In 1995, just under 50 per cent of its business was with ICL, for which it competed in the same manner as all other potential service suppliers. The range of technical services it provides to clients is illustrated in Figure 4.5 where it can be seen that D2D has the skills and tools to support activities at every stage in the development of a new electronics product. As such, D2D supplies services in the areas of product design and manufacture; printed circuit board (PCB) fabrication; PCB assembly; system assembly and test; supply, software and documentation services; repair, refurbishment and recycling. These services which follow the various stages in a new product realization can either be carried out directly by D2D or can be provided as general support for client companies to carry out these activities themselves. In practice, there is usually a mix of these two approaches depending on where the precise capabilities of the client company reside.

The company is a major player in the European market and has the ambition to become the largest provider of electronics manufacturing services in this geographic area. Its mission statement is very precise on this issue:

To become Europe's leading electronics manufacturing services

110

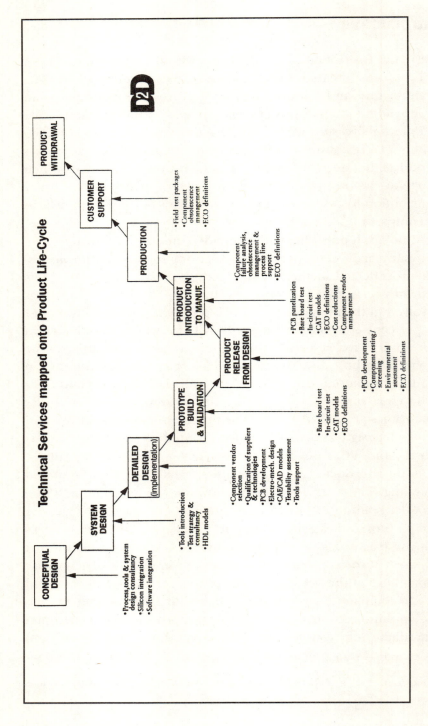

Figure 4.5 **Technical services provided by D2D**

company, in our chosen markets, by providing products and services world-wide that exceed our customers' expectations.

D2D's origins were in the manufacture of personal computers, workstations, client server systems, mainframe computers and other networked computer systems. The company retains a strong lead in this area of electronics manufacture. However, as more of its business has been derived from non-ICL clients it has become involved in the development of a much wider range of products, including mobile phones, mass spectrometers, cash dispensing machines, supermarket checkout machines, head-up displays and similar products.

Competitive criteria

The major market pressures influencing the activities of D2D concern those typical of the electronics industry: a requirement to shorten time to market whilst simultaneously reducing cost and ensuring defect-free products. Typical time to market requirements for mainframe computers have been seen to reduce from five years to three years in the period since 1980, whilst capabilities and complexity have increased enormously. The reduction for personal computers is even more dramatic, with times to market for these products currently standing at three or four months, a situation highlighted in Figure 4.6. In addition to these market pressures, there also exists the need for increased flexibility of batch sizes and a requirement to meet environmental standards associated with safety, electromagnetic compatibility and recycling of components.

In parallel to the changing market conditions experienced by D2D, the product technology is advancing equally rapidly. As is found throughout the electronics industry there is a continuous move towards further miniaturization. This is driven by the requirement for increased functionality for a given component size resulting in an increasing number of connecting pins on the main printed circuit board components. In consequence, there has been a continuous reduction in the pin pitch to a stage where

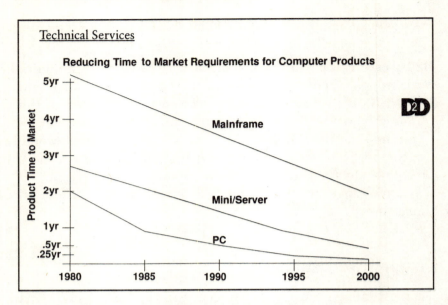

Figure 4.6   **The reduction in time to market of computer products**

there are now approximately 6 pins per millimetre in standard components. This compares with 2 pins per millimetre five years ago and 4 pins per centimetre fifteen years ago. In manufacturing terms this has led to the requirement for continuing developments in the area of interconnectivity between the various components. With pins being so small the challenge is to ensure that manufacturing processes are capable of reliably populating printed circuit boards with such miniaturized components in order to produce a fault-free board.

D2D emphasizes its capabilities in the areas of the design process, computer-based design tools, component knowledge and strong links to component manufacturers. It is capable of supplying a front-end design service for many product domains and can provide project teams which will take the initial concept through to manufacture, including a sophisticated prototyping capability. It has wide expertise in the areas of conformance criteria, electromagnetic compatibility, shock and vibration testing and legislation regarding product standards. Finally, it provides logistics expertise and facilities to ensure that the product can be rapidly and efficiently distributed to the final customer. This distribution capability additionally differentiates D2D from its service supplier competitors in that it will organize suitable packing, storage and transportation to, and within, any country in the world.

One of the strategic responses of D2D to changes in the market conditions and the product technology has been to apply a regime of continuous improvement to all the company processes. This has been realized, in part, by the investment in computer-based tools to aid in the product development and in information management by maximizing communication both within D2D and with its customers and through automating many of the engineering processes. There has been a steady growth in the utilization and sophistication of the information technology infrastructure to ensure cooperative working through the rapid dissemination of information. Thus the approach to implementing concurrent engineering within D2D can be seen to have been realized through technological developments in the use and coordination of knowledge and information.

The client server project

An example of concurrent engineering as practised by D2D was illustrated by the design and manufacture services provided to one customer who was developing a new range of products. The product, a UNIX file server, is a computer system in which software is located and available to be downloaded on request to networked PC workstations in a client server architecture. There is a rapidly growing world market for client server products as more users realize the benefits available from convenient availability of software across networks, often geographically dispersed.

The market for client servers is highly competitive, with product life-cycles being in the order of approximately one year. With electronics capabilities increasing so quickly the product rapidly loses competitiveness, often requiring a completely new design, rather than a simple upgrade, to compete successfully. Therefore a necessary capability that D2D had to demonstrate before being awarded the contract to provide manufacturing services was that it could support a significant reduction in time to market relative to previous projects. The target in this instance was for the first client servers to be available just six months after commencement of the project, whereas previous best times had been twelve months. In addition, the range of products would be larger than previously attempted, and component miniaturization would be further advanced.

These requirements implied that D2D would have to liaise very closely with the customers from the very earliest stages. Due to the complexity of the product and the design resource required to complete the whole range, several customer sites were involved, not only within the UK. Two of the client servers were to be designed in Sweden and the other four were to be split between two UK design sites. In addition, the chip sets differed across the range of client servers, with SPARC being used for the higher performance UNIX units and Intel for lower performance UNIX servers and PC workstations. To ensure commonality of approach in the design process it was necessary for there to be constant interaction between design sites and for the manufacturing service supplier to be equally well integrated into the design process. The manufacturer of the client server would have to be able to liaise with several design sites, provide the necessary input to ensure manufacturability of the product and plan the final manufacture and distribution.

Financial considerations

Since D2D supplies a full range of services to cover activities throughout the development process, the project can be scoped early and refined as it progresses. Whilst there would normally be several stages in a project where financial negotiations take place, the early scoping significantly reduces the likelihood of unforeseen cost implications. Once the design has been assessed in terms of both functionality and manufacturability then the actual cost of manufacturing can be specified.

D2D operates an 'open book' policy whereby client companies

can fully identify all D2D costs within any quotation. Contracts and quotation teams are in regular contact with customers throughout the progression of the project. It is possible for projects to be halted by the customer at any stage in the process if the financial viability of the product is called into question. This is, however, highly unusual and if it does occur is caused by changing market requirements being imposed after the start of the project. Additionally, the option is always available for customers to contract out certain activities to other service suppliers. Whilst this may provide apparent short-term benefits, in practice it proves to be a very dubious way of reducing costs. The need to establish new relationships and audit capabilities inevitably results in delays and unexpected increases in costs if service suppliers are switched in mid-project. Undoubtedly the most mutually beneficial arrangement is based on long-term relationships which include understanding and trust between customer and service provider.

Order-winning criteria

D2D was awarded the contract based on its ability to match the above set of requirements. It demonstrated that the combination of its capacity to merge into the project via its various information technology communication channels and its competence in terms of design and manufacturing capabilities would provide its customers with the expertise in order to reduce time to market. It had a proven ability to contribute during the early stages of a concept design and demonstrated the beneficial effects that this could have on manufacturing the product. In that respect it provided the crucial aspect of concurrent engineering to the project – ensuring timely consideration of all downstream activities.

It is worth noting that compatible and efficient communication networks between customer and service supplier are no longer specific order-winning criteria. They are treated as 'qualifiers' which need to be available before the service supplier is considered as a serious contender for a contract. Competitor service providers are now capable of rapidly establishing communication links to support design and it remains the quality of service that determines whether a contract is awarded. D2D was awarded the contract by initially demonstrating its information technology capabilities, but it crucially demonstrated its capabilities and experience in terms of providing the vital design embodiment and manufacturing engineering knowledge.

Project progression

It was the customer's responsibility to progress the basic functional design of the client servers. The customer employed several design sites who themselves were communicating on a very regular basis to coordinate different elements of the product design. In this situation D2D had to provide coordinated support to several of the customer sites. All support was provided from remote sites so communications had to be confirmed at the earliest stages in the project. Following that, it was necessary for D2D to employ compatible design tools in order to ensure no ambiguity in the generated data between customer and service supplier.

The support that D2D provided to the project is shown graphically in Figure 4.7. The involvement of D2D engineering services commenced at the initial conceptual design phase where CAE tools were utilized to help the client develop the most suitable design whilst taking into account all aspects of manufacturability. It is well accepted in D2D that this early stage of design is where the majority of manufacturing cost is committed and considerable activity was exercised at this stage. High speed data communication links ensured that concepts could be remotely viewed on screen by all contributing designers, with designs being developed concurrently in terms of both functionality and manufacturability. A complex design developed on one customer site could be viewed by D2D personnel within a matter of minutes to ensure that no excessive manufacturing costs were being unnecessarily designed in.

Once the concepts were firmed up then D2D moved on to

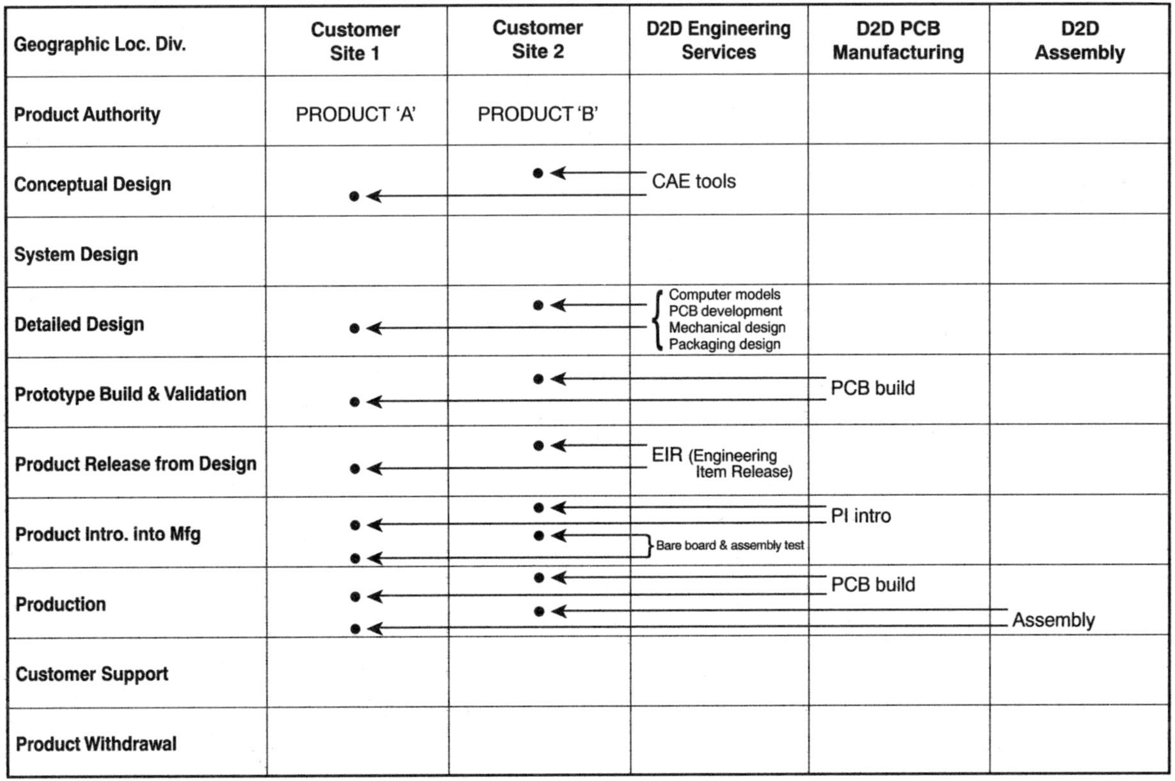

Geographic Loc. Div.	Customer Site 1	Customer Site 2	D2D Engineering Services	D2D PCB Manufacturing	D2D Assembly
Product Authority	PRODUCT 'A'	PRODUCT 'B'			
Conceptual Design	●	●	CAE tools		
System Design					
Detailed Design	●	●	Computer models PCB development Mechanical design Packaging design		
Prototype Build & Validation	●	●		PCB build	
Product Release from Design	●	●	EIR (Engineering Item Release)		
Product Intro. into Mfg	● ●	● ●	Bare board & assembly test	PI intro	
Production	● ●	● ●		PCB build	Assembly
Customer Support					
Product Withdrawal					

Figure 4.7 **The activities of D2D on the client server project**

supply services in the area of simulation, PCB development, mechanical design and packaging design. Validation of the product design was ensured through D2D's engineering database which carries a complete description of all products previously designed and all components that D2D have validated as being suitable for inclusion in products. On completion of validation the client server designs were formally released from design and a firm quote developed for the cost of manufacture. At this stage D2D manufacturing and assembly teams took over from engineering services to complete final manufacture.

D2D supplied design advice and design data from compatible software tools over the network to the customer design teams during several stages in the process. They provided primary input in terms of design for manufacture during the concept design phase, but they also provided customer sites with support in the most effective use of software design tools. It is policy to ensure that client design software is replicated within D2D to ensure full compatibility of data. D2D engineering services were heavily involved in conceptual design, component modelling and in the printed circuit board development. The matrix of activities and services supplied by D2D included engineering services providing compatible CAE data to various design sites, whilst PCB manufacturing provided services to validate the design, offer manufacturing support and commence order of components in order to schedule in the manufacturing processes.

IT support infrastructure

The rapid provision of expertise to its customers is one of D2D's major differentiators as a service supplier and was demonstrated in the client server project. In order to optimize this provision D2D has developed a very extensive support infrastructure based on information technology. At one level it has the capability for rapid communications between all its sites and with its client customers. Its basic method of communication is via its electronic mail system. This is an integrated office system providing at the local level word processing and spreadsheets etc., but at the wider level internal communications and a gateway into the Internet. Customers, suppliers, partners and academic institutions throughout the world are therefore all available over this network. It provides for standard text transmission and has now become one of the accepted methods for communication within D2D. In effect, colleagues and customers located in other UK sites, in Europe or in America can interact on projects in much the same way as co-located employees.

Increasing use has also been made of videoconferencing in the past three years. The significant advantage of this form of communication was given as the 'eyeball to eyeball' contact. Whilst the electronic mail system has been seen as a very significant step forward in terms of data exchange, it is recognized that the visual interaction available through videoconferencing has a significant role to play where complex issues require debate.

Virtual teams

The effect of this communications infrastructure is that when multidisciplinary teams are set up within D2D to coordinate new product introductions, they are not necessarily co-located and except for the key players are rarely full time on any one project. Some teams are made up of members dedicated to a single project but frequently individual engineers are, in fact, members of several different teams. The project team becomes a virtual team and rarely if ever will all its members come together physically. Only the project manager, who is involved on a full-time basis, will travel extensively to meet project team members at both the customer and D2D sites.

Project management

A networked project management package is used as the standard software tool for monitoring project progression. Team members have access at all times to identify their own individual set of actions and how these relate to the overall progress of the project. On some occasions, such as the client server project, the customer conducts the primary planning, monitoring and control of the project's progression, with D2D interfacing into the project management software. On other occasions D2D will have overall responsibility for planning and control once the client has defined key dates and stages.

In addition to the electronic mail system a high speed data link has been provided as a link between key D2D sites. However, whilst this ensures a higher transmission rate it is the fact that it is dedicated to D2D that is of primary importance. This is in direct contrast to electronic mail. With the Internet becoming so congested its response can no longer be relied on and at peak times it almost 'seizes up'. A dedicated high speed data link is not only fast but its speed is guaranteed at all times of day or night. With this facility it becomes possible to employ a team of design engineers, located on different sites, to optimize any given design. This ability enhances the potential for design engineers at the concept phase to consult with colleagues further downstream. Since turnround time in providing input to concept designs is so short and the benefits so obvious, then this element of concurrent engineering becomes a natural part of the process.

Engineering data management

The approach within D2D has been to develop a fully integrated engineering data management (EDM) environment facilitating data interchange throughout the design to manufacture process. The relationship of the various components of this environment is illustrated in Figure 4.8. This automated approach to engineering data management was essential if D2D was to achieve its objective of enabling distributed individuals to form the virtual team environment. Once all of the data links had been established and verified then the physical location of any individual designer was irrelevant to the progress of the project.

In principle the structure of the system is straightforward. An engineering parts and structure database holds all information on past products. This information is directly available for manufacturing planning via the design engineering to

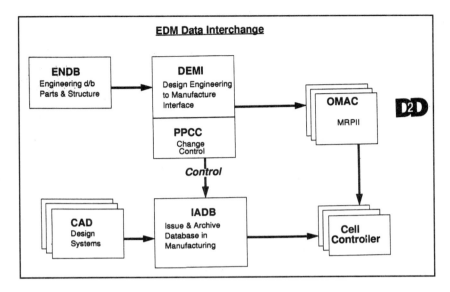

Figure 4.8   **EDM data interchange**

manufacture interface (DEMI). For new products or for product changes this data is available through the change control interface, product planning change control (PPCC), and can be used to create necessary data within the issue and archive database in manufacturing (IADB) for individual manufacturing cell control.

The reliable operation of this approach to data management is fundamentally dependent on the integrity and validity of the data held within the various databases. The data held within the engineering database (ENDB) is particularly prone to being changed and updated. Components are continuously being revised or withdrawn by manufacturers, with notifications of such changes being communicated to D2D at the rate of over 100 per month. Under these circumstances it is not surprising that D2D operates a significant procurement engineering group comprising approximately 20 to 30 engineers at any one time. New components must be evaluated before acceptance, the database updated, and project engineers informed if component changes are imminent for existing products or for products currently under development.

Computer aided engineering

In order to ensure compatibility with clients' in-house CAE software, D2D has available the three most popular standard suites of software for design through to manufacture of electronic products. Since each of the suites has its own strengths and weaknesses, it is unlikely that standardization will occur in this area for some time to come. Whilst data can be exchanged between the different CAE systems, the requirement for D2D to be completely integrated into a client's design process led to the strategy of operating three of the major systems.

These comprehensive design to manufacture tool sets are supported by more specific 'point tools'. They provide capabilities in a range of areas, including development of programmable logic devices, simulation of performance and timing characteristics, generation of PCB board panelization data and bare board test information. In addition, data conversion software is available to adapt data from those customers using CAE software which may no longer be fully supported and/or has a very small user base.

Engineering database

As far back as 1976 the engineering database (ENDB) was created to be the standard corporate component library. It has grown steadily over the years to match the increasing variety and complexity of the electronics products with which D2D is involved. The database holds hierarchical data in a typical bills of materials-type structure and can therefore be used to provide top-down information such as subassemblies to be reused in new designs. Alternatively it can be used bottom up, when it is necessary to identify all products which are affected by a component specification change. For whichever purpose the database is used, it is the integrity of the data which remains central to its effectiveness.

Design engineering and manufacture interface

The interface between manufacturing planning and the

engineering database is straightforward for existing products, subassemblies and components. A direct link is established via the DEMI interface. However, where new designs are being developed or changes are required to existing products then there is a need for change control. This is achieved via the Product Planning Change Control package (PPCC) which provides a validation service used in all designs to ensure that data is complete and valid to the level of detail required. As each phase in the design process is reached then the design is checked by the PPCC to ensure that no data is missing. Once this has been satisfactorily achieved then the next stage of design detail will commence. On completion of the process the data is utilized in the next stage of providing manufacturing information via the Issue and Archive Database (IADB).

Issue and archive database

In 1987 the IADB was introduced to act as a buffer between engineering and manufacturing. Design data which has been validated in terms of completeness and integrity by PPCC is transmitted to manufacturing via the IADB. When this database was initiated the concept had been that final validation of the data would be ensured through manufacturing trial runs. Any problems associated with manufacturability would become apparent at this stage and the design could be changed. Final validated data could then be approved by manufacturing in the IADB.

However, so rapidly has the world of electronics manufacturing changed that the validation process is now almost completely the reverse of that intended. With flexible production lines replacing fixed facilities the correct arrangement of machines and processes requires physical validation. As more sophisticated software packages become available, the manufacturability of a product carries a higher level of confidence. The combination of these two factors has seen a significant shift away from trial runs being used to validate only the data towards the data being used to validate the complete production facilities.

Conclusions

D2D is a service provider in that it supplies extensive design and manufacturing support to client customers. All the products in which it is involved are initiated by original equipment manufacturers. This means that concurrent engineering has to take place across the divide between customer and supplier. Design and manufacturing will often be physically separated and any implementation of concurrent engineering has to take this factor into account. The extensive information technology infrastructure provided by D2D allows quick and easy communication between everyone involved in the development process and removes the usual concurrent engineering requirement for co-location of teams.

The extensive component database and the software validation programs ensure that each stage of the design process is complete before the next one is started. Efficient engineering data management provides D2D with the capability to develop new products and to optimize manufacturing processes such that they are constantly reducing time to market. The future understandably lies in the greater integration of D2D with customers through the application of IT to ensure an even more rapid delivery of services.

Tools and Techniques for Efficient Product Development

IBM UK, Havant

Stuart Marshall, Sa'ad Medhat, Graeme Pratt and Jim Rook

Introduction

A concurrent engineering approach has been adopted in storage
subsystems development at IBM UK, Havant, in order to maintain
a competitive advantage. The company develops and
manufactures disk drives and associated technology for IBM's
RISC System/6000 workstation products. The products are
industry leaders in performance, which is partly achieved by the
use of serial link interface technology between the disk drives,
array controller and computer adaptor. This serial interface, now
in its second generation as an industry standard with improved
performance and connectivity, is known as serial storage
architecture (SSA).[1]

Open architecture environments, containing design
automation and project management tools together with
integrative databases, assist in the monitoring of product
processes. They enhance the communication of innovations,
requirements and definition for the product development cycle
as a whole. Any failure to integrate and disseminate leads to

conflict and a subsequent breakdown in both communication and direction.[2]

The availability of a common database infrastructure, accessible by all departments within the organization in the form they require, has been the aim for IBM.

Organization

The storage subsystems business unit is organized as shown in Figure 4.9. The three major functions of a product team are represented in one group, these being development, manufacturing and marketing. The structure of the development organization is a matrix of product and skill groups with clear boundaries of responsibility. This structure was established to address the increase in the number of projects and the need for reduced cycle time. Figure 4.10 shows the profile of cycle time and the number of products over the last few years.

The organization of project teams at IBM is a matrix style, as shown in Figure 4.11. A number of different skills combinations may be required for the various projects and so multidisciplinary development teams are formed. Projects are operated on and delivered using empowered and self-directed teamwork, with some team members being the champions of their skill area.

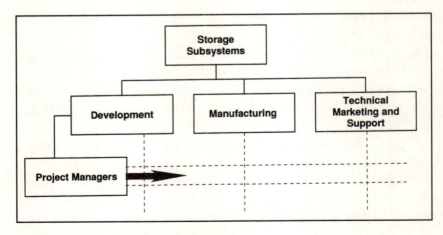

Figure 4.9   **Storage subsystems organization**

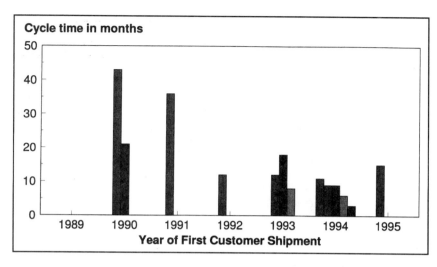

Figure 4.10 **Development cycle time**

The ability to work effectively as a member of a team is critical. Using multidisciplinary teams is not equivalent to forming committees where members often delay decision making. Instead, design teams get faster actions through early anticipation, identification and solutions to problems.[3]

Skill managers have the development team reporting directly to them and have the responsibility for developing the skills of their department. They also make commitments to the project managers to supply skills to support the product development schedules. Any resource conflicts are resolved by the development manager who can prioritize the work and authorize recruitment if necessary.

Product development structure

Responsibility for a product's profit and loss account resides with the product manager. As the product manager usually has a family of products to manage, he or she often delegates business responsibility for one product to a programme manager. The programme manager is chairperson of the product management team, which contains representatives from all areas of the

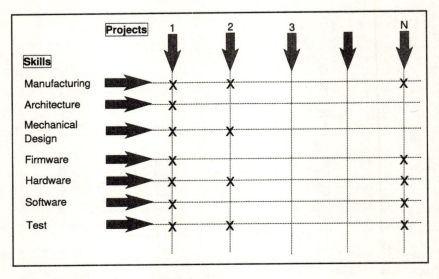

Figure 4.11   **Matrix structure for project teams**

company. Typically these would include representatives from:

● the development team
● business planning
● worldwide marketing
● manufacturing
● service.

The typical project management structure is shown in Figure 4.12. The product launch team (PLT) and the product development team (PDT) both report to the product management team (PMT). The PDT represents development activity and consists of representatives from American sites as well as Havant. Development, test, manufacturing, service and support status are addressed at meetings of the PDT. The PLT represents worldwide marketing, planning and finance. Pricing, product positioning, competitive analysis, education, customer fulfilment and technical support are addressed at meetings of the PLT. Meetings for the PMT, PDT and PLT usually occur in America in San Jose, Tucson and Austin.

Local project control and technical teams exist in Havant and

129

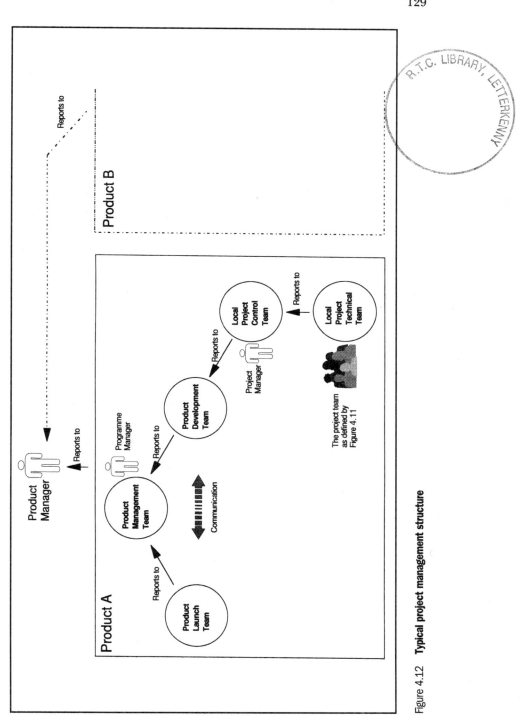

Figure 4.12 **Typical project management structure**

meetings occur in the form of project technical meetings (PTMs) and project control meetings (PCMs). PTMs are particularly useful in highlighting areas of critical development activity so that resources can be reallocated more usefully. PTMs report to PCMs where any important issues are highlighted. Whereas functional staff attend PTMs, functional management attends PCMs where development issues are raised and the project's status is evaluated. Typically, such a meeting will highlight issues that are preventing entry to the next development phase. In addition, PCM issues will be reported to the PDT.

The product management team meets at regular intervals throughout the development cycle. The programme manager is responsible for the business process, which has the following 'checkpoints':

- initial business proposal
- product announcement
- first customer shipment.

These checkpoints provide a progressive evaluation of the product business outlook as well as a thorough technical assessment as development proceeds. The development team is represented by the project manager whose task is to deliver the product into volume production on the agreed schedule. The project manager must create and execute implementation plans in order to assure that the completed product meets cost and expense commitments. He or she negotiates resource commitments with skill managers to develop and test the product, as well as coordinating all the business checkpoint tasks assigned by the programme manager.

Product development process

The increasing pressures of time, cost and quality demands have been a key driver for IBM to look at its business and the complete engineering development process. The factors contributing to these pressures include:

- competition from companies that are fast to the market with new products
- demands for better performance and 'value for money' from customers
- more international collaborative projects
- a need to replace aging systems
- a need to be more flexible and responsive to market opportunities.

A review of the characteristics of the development process revealed activities that consume major amounts of time and money. These offer the opportunity for improvements through the reduction of:

- design lead-time
- engineering changes
- direct and indirect costs
- the amount of paperwork
- the transition time from design to manufacture
- data management, searching and document handling times.

An overview of the typical product development process is shown in Figure 4.13. At IBM, product requirements are distilled into a set of product objectives by the project manager. These objectives form the basis of a resource-sizing activity and give the programme manager an understanding of the tasks involved. The project manager provides an outline schedule to the skill managers and requests a detailed resource- and expense-sizing from them. This is done for all projects and the technical planning manager consolidates the entire plan in order to assess the overall development expense budget for the organization. Work then starts on a high level design and functional specification which is a description of the product. It is reviewed to ascertain that it meets the requirements statement. It is also at about this time that the first cost estimate is done for consolidation into the initial business proposal. The marketing representatives provide sales volume estimates and a market price which is derived from competitive analysis. We now have

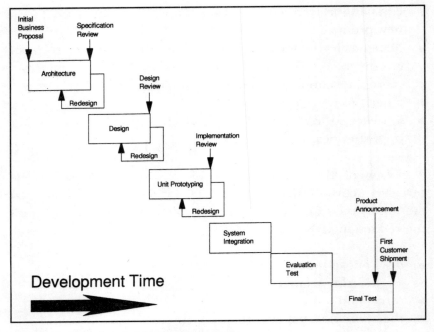

Figure 4.13 **Typical product development process**

the core ingredients for a business analysis:

$$Profit = Volume \times (Price - Manufacturing\ Cost) - Expenses$$

where *Expenses* are the sum of marketing, development, warranty and service expenses plus corporate allocations that fund research.

Plans for marketing, test, technology, manufacturing and service are also formed at this early stage of the development cycle to make sure that it makes good business sense to proceed with the project.

The new product development team includes hardware and software engineers, who work in conjunction with test personnel at the earliest opportunity, in order to allow testing to be efficient and meaningful. This ensures the discovery of problems at the earliest possible stage of development, so that the final testing can be performed with the minimum of defects. Therefore much effort is put into reviews before hardware and software implementation begins.

A 'specification review' which looks at the high level design concepts and a 'design review' which looks at pre-implementation design specifications are both carried out before any work is implemented. These reviews are placed in the initial project schedule and act as evaluation checkpoints. Participants of the reviews include the following members:

- author – the specifications writer;
- moderator – ensures adequate preparation before the meeting, maintains discipline during the meeting and draws satisfactory conclusions at the meeting;
- reader – any attendee, except the author, to read each section of the material;
- reviewers (of which at least two are required) – query and probe the subject area.

Representatives with knowledge of the particular aspect of a project under review will attend these working meetings. An 'implementation review' will also occur after a design has been implemented.

The mechanical design team operates a 'fast path' approach where development, manufacturing and procurement engineers sit together to bring representative packaging models into test at the earliest opportunity. They operate two databases, one containing the description of models being built by manufacturing and another containing the latest ideas from development. A review process transfers the development parts to the manufacturing database subject to stability and availability of parts. Shortly before the product is in full production, the manufacturing database is transferred to the formal release process which enables products to be manufactured in any IBM plant. Reviews will occur for the mechanical design in a similar fashion to the hardware and software aspects of a project.

Simulation is performed at chip, card and system level. The micro code is also simulated during the design code and unit test process before being merged with models of the hardware in the card and system test. In general, fault isolation becomes more difficult the higher the level of test that is performed. System tests do, however, represent an environment that is closer to the final application. The system integration phase, where hardware and software are brought together for testing, will highlight any major compatibility problems.

The product planner drafts the formal announcement document which states the full ordering structure, price and availability date. Before product announcement, an 'announce readiness review' is carried out to check that all aspects of a new product introduction have been covered. For example, the order process is checked, a search against patents is made, a plan for educating service personnel is made and marketing prepares the launch material.

With many products, customer involvement starts at the design stage and prototype units are made available to enable early evaluation and feedback on their performance in real application environments.

Engineering data management

IBM's commitment to communication is based on its worldwide internal electronic mail system. Virtually all development data and documentation can be transmitted throughout the system. Communication is further enhanced via local area network (LAN) applications that share common data. This enables schedules, engineering and financial data to be accessed by anyone with access to the LAN.

Problems of a more abstract nature that might relate to a general problem encountered in, for example, programming methods can be discussed in on-line forums. This ensures that a problem that might otherwise be ignored, or just discussed between a couple of people, becomes visible. Another person discovering a problem can search for any previous incidence of the problem and gain a much better insight into its cause and resolution.

With many projects running concurrently, there is a need for tight control of schedule, as any delay of a critical resource can have an impact on many projects. A PC LAN-based planning tool is used and the skill managers are responsible for tracking the progress of their deliverables into the main product schedule. The project manager maintains the overall schedule and holds frequent regular meetings with representatives from the whole team to review progress against the plan. There is also a weekly forum for all skill and project managers to discuss unresolved issues.

The project managers have formed a 'project office' where software tools for managing the information on progress can be developed. They make a risk assessment against their schedule milestones on a monthly basis. This is a prerequisite for the development report at the product management team meeting.

A recently introduced company-wide software tool, known as configuration management version control or CMVC, is used during new product development at IBM. It is designed for use in a networked environment, where software located on a server controls all data throughout the development cycle. Workstations running 'client' software are used to access the information on the server, allowing relevant project data to be worked with by team members. Software and firmware files under development are maintained in a file system and are managed by a version control system. A relational database on the server maintains all other development data. The organization of projects within CMVC is hierarchical in nature and an example of this hierarchy is shown in Figure 4.14.

As can be seen from Figure 4.14, the top level component defines the project. A hierarchy is then developed to reflect each constituent of the product, in this case hardware, software, firmware and architecture. These components are expanded as appropriate, for example modules within firmware. The hierarchy formally defines the areas of product development and each component serves as a storage space to hold specific data. There are no inter-constituent connections.

The problem log is maintained on CMVC and is accessible to any employee with a user ID, to record details of problems found during development. All discourse about the problem is accessible to any user with the required authority. The problems are categorized according to their severity, age, target date,

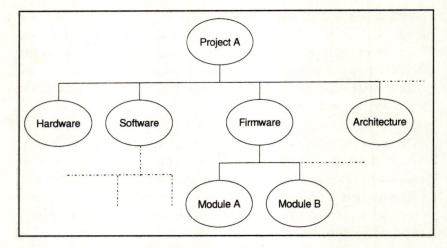

Figure 4.14 **Configuration management version control – typical structure**

owner, component and originator. Severities are classified from one to four, with one being the most severe.

Since all interested parties have access to this discourse, a very wide range of input is possible so that, for example, somebody generating documentation can be kept up to date about a design change that will affect him or her. The problem log becomes very powerful as a development tool during test phases. When a developer discovers a problem in a product during a test, it is raised on CMVC. The circumstances under which the problem was discovered and a code dump that will reveal the exact nature of the problem can be recorded. External suppliers also have access to the problem log so that problems that either affect them, or are affected by them, can be dealt with in a similar fashion.

The very significant documentation and clerical task workload associated with change and configuration management becomes a 'background task' automatically managed by the system. The system also provides full traceability by maintaining a complete history of all the changes. These actions undertaken by the project team as they create, use and modify the product data generate the required change history information.[4]

Defect analysis and prediction

Methods used previously for monitoring the product development process have in certain cases included the analysis and prediction of problems found. However, these methods do not take a holistic approach in terms of process analysis of new product development where many engineering disciplines can converge (e.g. computer systems). Moreover, they tend to concentrate on one aspect of the product, usually the software element. Much more emphasis is being placed on software within IBM's electronic systems, because software can be more easily and rapidly changed than hardware and can be used to fix either hardware or software problems.

When problems occur during new product development, they are documented and classified as defects. A large proportion of software programming expense can be attributed to the detection and removal of these defects, and the most cost-effective removal methods are those that eliminate the defects as early in the development cycle as possible. Various metrics have been proposed relating to software defects, to plan, control and evaluate the software development process, and this enables data to be collected and analysed in a meaningful way.[5, 6]

A model to predict the number of defects during development will help to provide a more efficient new product introduction and may provide an assessment of when to end testing of a product economically. This is useful because testing is expensive and needs to be optimized to provide a balance between test coverage and costs associated with the testing. When the rate of defect discovery starts to decrease, and providing that the project's progress is stable (e.g. there are no development problems that are bringing testing to a halt), then the test phase can be deemed close to completion. If the project schedule shows more testing resource than is necessary, then there may be scope for reductions in time allocated to testing. The outcome is a more efficient development process and improved product quality.

Reliability growth models have been used in the prediction of software defects.[7, 8] Such models require the use of data obtained relatively early in the software development life-cycle, to provide reasonable initial estimates of the quality of an evolving software system.

At IBM, a log is made of all defects found during new product development on CMVC to give accurate and comprehensive statistics for defect analysis and prediction. The most important processes monitored in the development cycle with regard to defect analysis are summarized below:

- specification review (SR);
- design review (DR);
- implementation review (IR);
- hardware and software integration;
- evaluation test (ET) – a formal prototype evaluation. In addition to classical tests such as EMC, vibration etc., tests of functionality, compatibility, software, firmware, hardware and error recovery procedures are included;
- final test (FT) – evaluation of the product against specification using parts manufactured by the proposed manufacturing processes (in low volume). Fixes to problems found in evaluation test will be retested in this stage;
- first customer shipment (FCS) – the point when the product is officially first available to customers.

From the defect data available, estimates of future defect numbers for the current project can be made, with initial focus being placed on the firmware, as this yields by far the highest number of defects raised. An IBM internally developed software tool for making these estimates is being used as an interim measure for the prediction. An example of the curve generated by the tool is shown in Figure 4.15. It shows the significant differences in timescale between previous and current projects, illustrating the reduced time to market of the concurrently developed new product.

The IBM structured approach to team make-up, in conjunction with the recently improved defect logging tool, makes data relatively easy to obtain for use in any predictive modelling. This is due to the projects being divided hierarchically and to good communications. Another aspect of concurrent engineering that

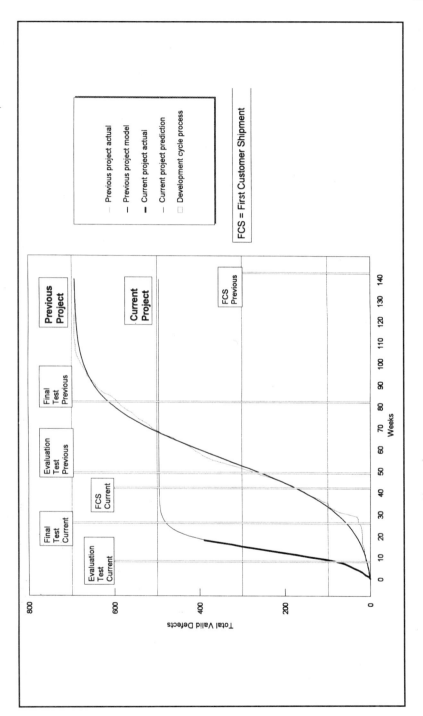

Figure 4.15 **Firmware defects from the start of integration**

helps in the production of predictions is the early involvement of downstream activities such as test in the development process. This allows defects to be found at the earliest possible stage and therefore defect data become readily available in the early project stages, to provide input to any model for predictions.

Conclusions

A summary of concurrent engineering activities at IBM Havant includes the following:

- team-building activities;
- high management visibility;
- the introduction of new teamworking practices;
- the cooperation of manufacturing and procurement via physical as well as electronic media (quality engineering techniques such as quality function deployment were used where necessary);
- hardware and software simulation;
- shared problem log;
- schedules available to all;
- early test involvement;
- communication enhancement workshop;
- full project reviews and meetings.

These processes and activities represent a significant and comprehensive effort to implement concurrent engineering. Communication has been regarded as essential, and this is particularly evident during the test phases where developers and testers work on the same part of a product, but may be physically located apart from one another. IBM also benefits from an extremely effective communications system which allows employees to communicate with each other immediately and globally. It is common for staff to have more contact with their project team than with the person seated next to them.

High management visibility and communication mean that all staff are fully aware of the current status of the project and of their role within it. The importance of physical as well as electronic communications has been recognized, and activities that enhance communication have subsequently been implemented to entice functional staff away from their work and into conversation.

One of the most effective development tools is the problem log, maintained on CMVC. Functional as well as management staff are able to understand all outstanding project issues, by having access to a common database where information on problems found during development is stored.

The philosophy of concurrent engineering arose out of the need to develop products competitively. One way in which this is achieved is by engineering defects out at the earliest possible stage of development. Enhanced control over the development process will become apparent and so ambitious schedules are much more likely to be met. Hence product schedules become a reliable timetable of processes and milestones throughout the project.

The defect prediction technique is currently being refined at IBM, by looking at enhancements to the models and evaluating the use of other tools. The use of additional techniques found in the concurrent engineering arena, such as quality function deployment (QFD), 'Taguchi' and producibility engineering, may also help in the task of producing accurate defect predictions during new product development.[9]

In a conventional development process, the discovery of defects at a late stage, possibly because there has been little concern towards designing for the complete life-cycle, dramatically increases the risk of a schedule being broken. The conventional process is very compartmentalized and does not include aspects such as design for test or design for manufacture at an early enough stage of a project. Finding defects in a late stage of a project may mean that a redesign is necessary. The impact of this will depend on the time that the defect was discovered and the work and resources required to fix it. In general, the later a defect is discovered, the worse the impact. With less control over the development process, the conventional development approach can be classed as unstable. Concurrent engineering leads to a more stable development process because of the enhanced control.[10]

Lessons learnt

There is relentless pressure from customers and competition to shorten development cycles, to produce lower cost products for less expense and to accelerate time to market. To respond to these market pressures, development, manufacturing, marketing and after-sales service have to work as a team focused on the common goal of reaching the market with a quality product. The diversity of products requires resources to be shared across these teams and this can only be achieved by having strict development controls and careful resource management. The engineers and programmers have to be flexible to respond to technical problems and changes in requirements. The use of common tools and predictive defect analysis is essential to manage the schedule exposures. As always, it is the quality and commitment of the team that will bring a successful product to the market.

References

1   'Information Systems – Serial Storage Architecture SSA-PH (Transport Layer)', Document 989D – Revision 3, 5 January 1995, X3T10.1 ANSI Committee. Available from Global Engineering, 15 Inverness Way East, Englewood, Colorado, 80112-5704, USA.

2   Medhat, S. (1994) 'Profit in Parallel', *CADD Journal*, **14**, (3),11–14.

3   Carter, D. (1992) *Concurrent Engineering: The Product Development Environment for the 1990s*, Wokingham: Addison Wesley.

4   Stark, J. (1992) *Engineering Information Management Systems: Beyond CAD/CAM, to Concurrent Engineering Support*, New York: Van Nostrand Reinhold.

5   Chillarege, R. *et al.* (1992) 'Orthogonal defect classification – a concept for in-process measurements', *IEEE*

Transactions on Software Engineering, **18**, (11), November, 943–56.

6   Neal, M. (1991) 'Managing software quality through defect trend analysis', *Managing for Quality*, proceedings of the Project Management Institute annual seminar symposium, Dallas, Texas, 119–22.

7   Caruso, J. *et al.* (1991) 'Integrating prior knowledge with a software reliability growth model', *IEEE 13th International Conference on Software Engineering*, Austin, May, 238–45.

8   Ohba, M. (1984) 'Software Reliability Analysis Models', *IBM Journal of Research and Development*, **28**, (4), July, 428–43.

9   Medhat, S. (ed.) (1994) *Proceedings of the International Conference on Concurrent Engineering and Electronic Design Automation*, Bournemouth, April.

10 Medhat, S. (ed.) (1991) *Proceedings of the International Conference on Concurrent Engineering and Electronic Design Automation*, Bournemouth, March.

5
People Issues

Measurement Technology Ltd and Temco Ltd

In this chapter two case studies are presented which illustrate the importance of considering people issues when introducing concurrent engineering into a company. People issues are fundamental to creating an environment in which change is seen as advantageous and will therefore be supported by all company employees. Whilst Temco Ltd and Measurement Technology Ltd, the two companies providing the case studies, have also considered those issues discussed in the previous case studies, such as changing tools, structures and processes, they both report their own activities to include a significant change in the way people work.

The stimulus for change in the two companies can be related to the concepts of Efficiency and Radical Innovation in the concurrent engineering framework (see Figure 5.1). In that respect they are experiencing similar pressures to those described for Marconi Instruments Ltd and Lucas Aerospace Actuation Division in Chapter 3. However, the different responses described in this chapter reflect the culture and characteristics of these two relatively small companies.

Measurement Technology found that rapidly changing capabilities in data transmission meant that its intrinsic safety products required dramatic technological upgrades in order to remain competitive. In addition, the time to market of these new products would have to be significantly reduced as compared to the previous generation of products. Temco, on the other hand,

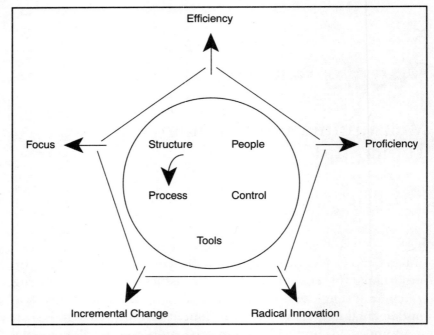

Figure 5.1 **The concurrent engineering framework**

which was suffering from poor supply of raw material, made the decision to develop a new plated strand from a specially created new alloy. With significant pressures on the company to cease using the old alloy, the development process had to provide guaranteed results in an acceptable timescale.

The process of change

One aspect which is striking in both of the case studies is the close collaboration between the two companies and academic institutions. Measurement Technology built on its links with the CIM Institute at Cranfield University, utilizing its knowledge of concur rent engineering implementation. Temco worked with the Department of Industrial Studies at Liverpool University which supplied support in tailoring implementation models of concurrent engineering. In both instances it remained the responsibility of the company to implement concurrent engineering, whilst the academic institution provided background knowledge and information.

There are clear differences in priority between the two companies resulting from basic and practical considerations. The issue to be addressed by Temco in sourcing a new alloy for its plated strand was critically dependent on the performance of personnel outside the direct control of the company. Temco was dependent on the performance of its supplier not only to deliver the required alloy, but prior to that, actually to develop the new material. The company was therefore reliant on the capabilities of a supplier's research and development activity. It is therefore not surprising that one of Temco's priorities in implementing concurrent engineering was to develop a relationship of mutual benefit between itself and the supplier based on personal interaction. The establishment of mutual strategic understanding was fundamental to ensuring on-time development of the new product.

In contrast, Measurement Technology had a slightly different set of priorities in implementing concurrent engineering. Whilst the development of improved relationships with suppliers was one of the priorities, it was not so strongly emphasized as in the case of Temco. For Measurement Technology the priorities were to develop its customer focus, improve manufacturing processes and stay ahead of the competition technically. Responding to this set of specific priorities, the activity carried out by Measurement Technology was more focused on the internal activities than was the case with Temco. Where external focus was required, such as learning from other organizations, then this was carried out. However, the diversity of focus between the two companies is quite clear.

Measurement Technology had a clear view that change would be best achieved through the people within the organization. Therefore organizational structures, processes and tools were not seen as the major issues to tackle. Instead, it was to be people's roles and responsibilities which would drive the implementation of concurrent engineering. Once these changes had been determined then the required changes to the development processes and organizational structure would follow. Training was an important feature of supporting people in their new roles and was seen as a strategic element of the implementation plan. This involved developing not only technical skills but also the interpersonal skills necessary to ensure closer cooperation both internally and with external customers and suppliers.

The approach taken by Temco was to consider the concepts of complexity of both the product and the product–user interface. The new product development process was designed to reflect the measured levels of complexity and it was utilized in a formal concurrent engineering implementation methodology. However, throughout the process of evaluation and implementation of concurrent engineering an ever present consideration was the maintenance of certain aspects of the company's culture and in particular its informal management style. Whilst there was a need to increase people's awareness of the importance of new product development, there was also the requirement to ensure that the active involvement and development of all staff remained a feature of the company.

Finally, it should be repeated that, although these two case studies have been brought together to emphasize the people issues in relation to concurrent engineering, they are both much broader than this one element. Both implementations followed a balanced approach which clearly looked at company requirements and developed the best solution for the specific situations. Not only did changes occur in terms of people issues, but also in relation to the processes, structures and, where and when appropriate, to the tools employed.

Changing the Organization through People

Measurement Technology Ltd

Wendy Bowden and Fiona Lettice

Company background

Measurement Technology Ltd (MTL) designs, manufactures and sells equipment for the control of industrial processes. It specializes in control equipment which is designed to be safe even when operated in the presence of explosive gas or dust mixes, such as those found in the petrochemical and pharmaceutical industries. The technology it uses is known as intrinsic safety (IS).

MTL was founded in 1971 by four design engineers who left during the restructuring of another instrument company, Kent Instruments in Luton. They started up their business in an old hat factory with family and friends comprising the workforce. Business developed rapidly in national and international markets and by 1988 the company was turning over £9.3 million with a profit before tax (PBT) of £2.3 million. In March 1988 the company was floated on the London Stock Exchange Unlisted Securities Market under the name of The MTL Instrument Group plc, with Measurement Technology Ltd as the largest subsidiary. Since then the company has sustained a steady growth rate of approximately 12 per cent real term growth per annum in both sales and PBT. In 1994 the group had a sales turnover of £30 million and a PBT of £9.4 million. It employs approximately 450 people worldwide.

Throughout its development the company has been enthusiastically international and early on began to establish a sales presence covering most of the world. In 1980 it opened its own sales company in America and today has companies in Australia, Belgium, Canada, France, Germany, Japan, The Netherlands and Singapore. In 1989 a joint venture company was established in India to manufacture two major product lines for local sales and to import other lines from Luton. In 50 other territories, MTL sells through a network of third-party distributors.

MTL is the world leader in intrinsic safety and currently employs approximately 340 employees at the Luton site. The company's products are typically small single circuit board electronic assemblies housed in a moulded plastic case. Some product lines are high volume (100 000 plus per year) low variety, and others, typically the more technically complex products, are lower volume (5000 to 10 000 per year) and tend to be made to customer order.

Why change?

A new managing director, Dr Graeme Philp, was appointed in late 1992 and quickly identified the need to improve the lead-time to market on new product developments. Intrinsic safety is a technically-based industry, and the fast pace of technology enhancement is driving a constant reduction in product life-cycle. The marketplace has tended to be quite conservative over the years, driven by the fact that the cost of instrumentation and control products tends to be low compared with the cost in terms of lost production if they fail. This has meant that the market requirements have changed slowly relative to other sectors in the electronics industry. Over the last few years, however, the pace of change has begun to increase with the ability to use digital data transmission to convey more information from the instrumentation to the user, therefore offering the opportunity to reduce running costs. MTL has had to get in shape to respond. An analysis of the business strengths and weaknesses and also of its core competences showed that three areas needed to be addressed to respond to these changing conditions:

- maintaining and increasing customer focus
- continuously improving manufacturing processes
- staying ahead of the competition technically by getting to market first.

The main conclusions were that customer focus and manufacturing processes were reasonably good, but could benefit from encouraging a continuous improvement ethos. Paradoxically, the company's technical leadership had led to products with long lives in the marketplace, which had removed the pressure for the continual development of new products. In light of the emerging market conditions, the area of product development needed to be reviewed and improved. New products were previously developed with a lead-time from concept to market of eighteen months to two years. The new target is for development lead-times of nine months for new products and six months for product improvements.

It was Graeme Philp's belief that it was through the people in the organization that change could best be achieved. Procedures, systems and information technology were not the primary issues, as motivated teams of people could make their own technology choices.

To introduce the changes, which would concentrate on the product development process but not ignore other areas of the business, a new role was created in MTL. As a result, Wendy Bowden joined the company from British Aerospace Regional Aircraft in March 1994, as organizational development consultant.

There was a very clear mandate to introduce concurrent engineering, so the first objective was to find out what was currently happening within the company. Wendy Bowden made it her business to get invited to all new product development meetings, spend time in manufacturing areas and generally speak to as many people as possible about the current state of play. The aims were to assess the effectiveness of the organization, in particular the effectiveness of product development projects, and to identify areas for improvement.

The main conclusions reached were as follows:

- Designers were leading product development projects regardless of leadership skill.
- Design teams were designing with little or no understanding of current manufacturing processes.
- Each function was responsible, but no one accountable, for overall project success.
- A high level of modifications followed each project into manufacturing.
- Key business processes were not clearly understood within the organization.
- The environment was comfortable and in general people felt very little need to change the way things were done.
- The common parts list was easily ignored.
- The total cost of manufacturing was high.
- Customer empathy was not on the agenda when designs were discussed (there was little focus on customer requirements).
- There was no training strategy.
- Learning from previous projects was not being utilized.

Introduction of concurrent engineering

Given these findings, a plan was produced for the introduction of concurrent engineering, which consisted of five key elements:

- roles and responsibilities
- project leader selection
- business process improvement (for all key processes identified)
- training
- learning from other organizations.

Roles and responsibilities

The first area of focus was the leadership of product development projects. The key question was, 'Who is accountable for the delivery of project objectives?' If this question cannot be answered quickly, then there is a grey area of accountability.

A workshop was set up with the managing director, senior management team and functional heads to address this issue. The task was to agree the roles and responsibilities of the project leaders and functional heads and finally what the project leaders and functional heads should expect of one another. The aim was to clear up accountability and to define the sort of behaviour that should be expected during product development projects. The workshop, which lasted one day, represented the launch of the concurrent engineering initiative within the company.

Before the workshop began, a set of roles and responsibilities was created for the project leaders and functional heads. These were put forward to be discussed and improved during the workshop. By the end of the day, the roles and responsibilities were clear, everybody agreed with them and it was felt that they could be used throughout the product development projects.

The output of the workshop is shown in Table 5.1. The result of the workshop was the first step to moving the organization from a largely functional reporting structure to the matrix organization shown in Figure 5.2.

For each new product, the project objectives are set by the business group manager. These objectives are largely driven by cost targets and the need to enter the market in the most opportune timeframe. Conflict between project and functional objectives has arisen, as there are many projects running at any one time. If the conflict cannot be resolved between the project leader and the functional head, then it is escalated to the managing director for resolution. He decides where the priority lies, and asks that the appropriate course of action is taken.

During this change to a matrix organization, it was the functional heads who felt particularly vulnerable as they had lost accountability for the achievement of project objectives. This can be difficult for some people to accept and needs to be carefully

Role of project leader
Achieving project objectives through the activities of the team
Responsibilities:
- To gather the resource through the resource holders
- To plan the project with a multifunctional approach
- To ensure monitoring and control mechanisms are in use
- To be the focal point for reporting
- To manage the achievement of project goals

Role of functional heads
Responsibilities:
- To ensure project leaders have recognized clout within the company
- To set direction and provide clear product development objectives for the team
- To provide advice and support to project leaders and team members, e.g.:
  - agreed resource
  - moral and professional support
  - provision of training and tools
- To provide timely decision making to the team
- To rearrange the management team meetings to incorporate project leaders

Expectations of project leader
- To motivate and coach the team
- To ensure functions are receiving information
- To develop a broader business picture, including all 'design fors' and to maintain team awareness of it
- To identify the training needs of team members and notify functional heads
- To be seen as a resource to the team to overcome blockages

Table 5.1 **Workshop output**

managed. It was important to stress that the functional heads would be responsible for creating an environment where multiple projects could be successful.

It was felt that there is 'no gain without pain', and you can be sure that if your organization is not feeling the pain of striving to improve, then your competition probably is!

Project leader selection

With this in place, the second task was to select project leaders for the competences necessary to achieve project objectives. Three

155

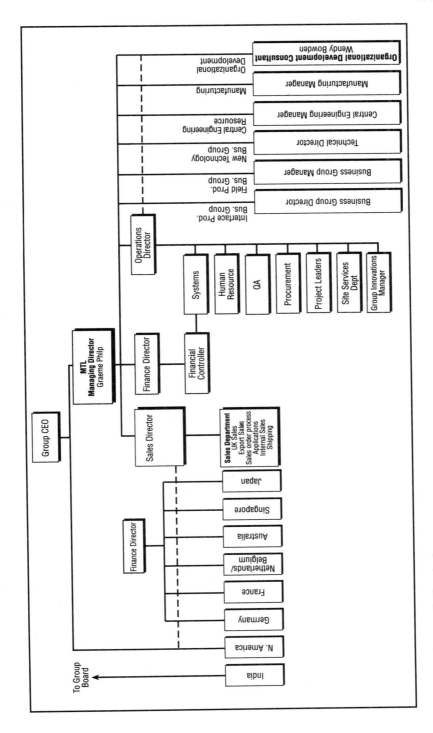

Figure 5.2 **New organization chart**

product development projects had been identified to drive the initial changes. Candidates were interviewed, and the most appropriate employees were offered the role of leader for each project, with the understanding that their responsibilities would be different to those of project leaders for previous projects.

The selection of suitable project leaders could be from anywhere within the organization. This was a definite shift away from design-driven teams, with design engineers leading them. In the past, project leaders had had a fairly narrow product design focus. As well as managing the projects, the project leaders were also making detailed design decisions and being actively involved in product development activities. Now, one of the major new product introduction teams has a production manager as project leader, and the other two are led by a mechanical engineer and a test and maintenance engineer. These new-style leaders do not become involved in detailed product development decisions. Instead, they concentrate on managing the team and project resources and take a broad view of the whole business, rather than just having an understanding of the function from which they have come. This change in project leader style and a selection of team members from all parts of the organization have resulted in truly multifunctional teams focused on a common goal.

What was good about this? It sent strong messages across the organization that product development projects would be managed differently in future. Positive results soon began to show as vendor lead-times reduced, team-based planning techniques allowed true involvement of team members, and prototyping was carried out by the team rather than just by the designers. This has been a positive step forward and has enabled feedback to occur earlier in the product development process about the manufacturability of the product.

During the early stages of the projects, a project leader was invited in from an aerospace company which has been practising concurrent engineering for some time, to talk to the three project leaders. This was beneficial in that the new project leaders were made to feel more comfortable with their responsibilities. They also felt that they were not the only ones to be going through such a change and were not alone in experiencing some of the problems that this type of change inevitably causes.

Another change has been the project team going out to visit customer sites to ensure that customer needs are being satisfied. These visits are hosted by the sales representative for that particular area, which is a good opportunity to bring the sales force into the multifunctional team. This is the first time that most of the team members have visited a customer site and started to understand how the customer uses the product that they help to develop.

When a customer site visit was being organized, a line technician was asked if he was going to be able to attend. His reply was, 'Do you mean I can go? I thought this was just for the top guys.' This is a reflection of how much of a change is occurring, and how more of the organization is now being involved and made to feel important in all product development activities.

Business process improvement

The third key element to be addressed was the improvement of the key business processes. The first task was to identify the key business processes, via facilitated discussion during a workshop session. The senior management team concluded on ten key processes. This may sound more like business process redesign than concurrent engineering, but how can product development projects be successful if current processes have been outgrown?

Ten process owners were nominated to lead the improvement activity. Some were from the senior management team, others were functional heads. Multifunctional teams were then formed to perform the process modelling of the 'as is' and 'to be' processes. As many of the processes overlap, some people are part of more than one team. The teams meet about once a fortnight to discuss progress and to identify and solve problems. Recently, a workshop was conducted to link all the processes together. Again, an expert was invited from another company with extensive experience of modelling business processes to help run the workshop. It was very useful to have an outside view on MTL and his presence provided a great stimulus to the business process improvement teams. A great deal was learnt from him in a short space of time, and the teams have subsequently felt more confident about the scope of the change that is achievable and feel braver and more able to tackle some of the tougher problems faced.

This exercise has without doubt reinforced the multifunctional approach, clarified how things are done and identified major improvement opportunities. Processes are very often historic rather than designed, and some parts of them are still performed even though they may no longer be necessary. This exercise can be time consuming and therefore must be taken seriously. A clear vision of why the business process improvement activity is being carried out is needed, and care has been taken to ensure that the benefits can be realized in the newly designed processes. A side benefit is that the process models developed will be used to help MTL obtain ISO 9001 certification in 1996. The activity has also helped to get more of the organization to buy in to the overall change process.

Some unexpected processes are being modelled, such as product termination. The initial workshop session highlighted that it is very unusual for products to be officially terminated within MTL. Generally, as long as a customer still requests a product it is manufactured, even though the product has been upgraded and is now easier and cheaper to manufacture. Another process which has been included is project evaluation and learning, which is key to the success of sustaining concurrent engineering practices. The inclusion of this process provides a means for carrying over lessons from one project to others currently running. More importantly, it will facilitate the launch and progress of subsequent product development projects. It will also help to standardize the product development process.

Training

The fourth element of the introduction plan was to develop the training strategy, ensuring that it supports the change and clearly defines future direction.

The first task was to put together some objectives for training which would form the base for any training programme design.

The objectives are:

- to develop customer empathy across the organization
- to cultivate team leader skills in order to maximize team performance
- to develop business awareness across the organization.

This sent a very clear message about where the organization is going and why people are attending training programmes. This could be summed up simply by saying, 'We are going to get closer to our customers, whilst working in teams, and make sure that our money-making machines make money faster.'

It is also important to ensure that training is translated into performance back in the workplace, as training costs us money. However, if you think education is expensive, try ignorance. There are also plans to run an open learning centre. This will be a couple of rooms set aside specifically to house training materials, information and self-learning packages. All employees will be encouraged to use the facilities provided, and will receive advice from members of the human resources department on training needs. Again, the training strategy will be much in evidence within the centre.

Learning from other organizations

The final part of the concurrent engineering introduction plan is learning from other organizations. This is key to promoting organizational learning and prevents having to reinvent the wheel. Many organizations have had different experiences of implementing concurrent engineering and managing change. These companies have been visited, whatever industry they are in, to understand what they have learnt as a result of their experiences. Given that the primary concern is with people and their behaviour during change, it does not matter what type of business you are in or what size of organization you have – the people issues will usually be the same.

MTL also participates in the CIM Institute Concurrent Engineering Forum, which is attended by a wealth of concurrent engineering practitioners, and provides a platform for excellent information sharing among the pioneers of British industry. Other companies have visited MTL and shared their own experiences and key learning points. The product development teams have also visited numerous customer sites. These activities have all contributed to providing inspiration and good ideas, and have helped to confirm that MTL is on the right track.

The main messages arising from other organizations' experiences are:

- Change must be supported from the top.
- Beware of senior managers paying lip service to the change.
- Be sure to support the change agents as they battle with the organization in their quest to get people to change.
- Celebrate success.

Conclusions

The biggest challenge of implementing concurrent engineering has been to tackle the people issues. In our experience, people are the key to driving change through the organization, and implementation success has rested on the ability of companies to provide a supportive environment, where multifunctional teams can work together with a common purpose. Technology and sophisticated tools may support business and product development improvement, but the people and attention to processes will determine business success or failure. Creating an effective team environment, identifying and improving key business processes and gradually changing people's behaviours and expectations have therefore been the primary focus of the change programme at MTL. Communication has been encouraged between departments within the organization and between the company and others. To improve product development performance, project management techniques and all major business processes, not just those directly involving product development, have been radically rethought. The shift to concurrent engineering has involved altering people's behaviour at all levels in the organization, and changing the way that work is done. A broad view of concurrent engineering has been taken to achieve this, undertaking any changes that will move MTL towards better business performance.

> The result can be seen clearly in the number of new products we are developing (achieving the highest levels of new product launches to date) – many of which are breaking totally new ground – and in the improved delivery times and services we are offering our customers.

> Graeme Philp, Managing Director of Measurement Technology.

What next?

The quest for improvement will never end. The intention is to instil an atmosphere of continuous improvement and an excitement to learn and develop both as an organization and as individuals. This process has been initiated through the changes and implementation process described in this case study. The next phases will involve actively cross-fertilizing experiences between projects. Project leaders from the product development teams are meeting regularly to share experiences. During their meetings they discuss concerns, share success stories and try to preempt resource conflicts between the teams by identifying when they are likely to occur and planning alternative courses of action. This and regular project progress reviews will help to facilitate organizational learning as teams spread throughout the organization.

Other key changes will involve altering management and measurement systems to reflect the new ways of working and to encourage team behaviour and attitudes. As well as reassessing product development process metrics and business performance metrics, employee appraisal systems will be reviewed. An individual's contributions to a team need to be reflected as well as their technical and personal competences in the individual reward system.

The aim is for the process improvement teams and the product development teams actively to drive many of the changes and their implementation. They will highlight the problem areas and appropriate courses of action can be jointly designed by business managers and the teams to keep the business moving forward.

This may all seem like common sense, but how much common sense is there about?

A Structured Methodology for Implementing Concurrent Engineering

Temco Ltd

Jane Burns, Ian Barclay and Jenny Poolton

Summary

An integrated concurrent engineering environment has been introduced into Temco Ltd, a subsidiary of BICC, through the application of a generic new product development (NPD) process. The implementation methodology was based on an approach suggested by several published sources, but was specifically adapted to fit the company's current and projected product portfolio, its core competences and its informal management style.

The approach of 'tailoring' the concurrent engineering environment to suit company requirements was facilitated through close industry–academe links. The structured assessment and evaluation method used in the implementation process forms the core of this case study description.

The success of the NPD implementation served to demonstrate the importance of concurrent engineering to Temco's future business environment. As a consequence, all new product initiatives are based around core competences within an established concurrent engineering environment. This both complements market developments and ensures reliable and rapid realization of new products.

Company background

Temco Ltd is the largest speciality wire manufacturer in Europe, with a product range that includes bare and electroplated wire, strands of basic wires, tinsellated and braided conductors, and enamelled wire strands. The company consists of two manufacturing units for the different product groups – one in Cinderford, Gloucestershire and the other in Nottingham. Business management, including sales, is based at Cinderford. The company, wholly owned by the Rod and Wire Operations of BICC Cables UK, employs approximately 140 people and has an annual sales turnover of about £10 million. The major markets addressed are high performance cabling for aerospace, datacommunications, telecommunications, and cabling and cordage for domestic heating appliances, loudspeakers and telephones. A wide variety of markets are served worldwide and 42 per cent of business is export. In 1991 Temco was awarded the Queen's Award for Export.

Plated strand

Following a core competences analysis carried out in 1991, plated strand was identified as one of the company's core products. Plated strand is purchased from Temco by cable makers who then supply the finished cable product to aircraft manufacturers. Its two main uses in this market are to carry electrical power and signals. Usage of electrical power cables per aircraft will increase in the future to facilitate advanced electrical control of flight mechanisms and to satisfy passenger demand for better comfort and entertainment facilities. In contrast, the need for signalling cabling will decrease in new aircraft as flight information management, currently realized through electronic instruments and cabling, is replaced by computer networks. However, there also exists a large refit market for both power and signals cabling as many airlines are refitting cabling systems to improve the performance and life of current aircraft.

The stimulus which drove Temco to source a new high performance alloy for its plated strand was a growing problem in obtaining good quality supplies of the original alloy which is recognized as being toxic. In 1992 a major programme was initiated to find a suitable non-toxic replacement from a reliable source. In significant contrast to previous practice, a strategic decision was made to focus business development on the core product of plated strand. This required a substantial evaluation of future market trends which could be influenced from a variety of sources, including cable makers, aerospace companies and product specifiers (Civil Aviation Authority, Ministry of Defence etc.). It was recognized that the product introduction process required significant change if the coordinated approach was to be successful. This implied the introduction of concurrent engineering.

A concurrent engineering assessment and evaluation framework

The next stage was the development of a framework that allowed Temco to 'tailor' its concurrent engineering introduction programme to produce an NPD environment that exactly suited its specific needs and circumstances. To do this, the methodology had to be capable of evaluating the company's concurrent engineering need and then allow a planned programme of development to be devised and implemented. In order to develop this framework, Temco and Liverpool University collaborated extensively. The input of the university was the knowledge base of a wide variety of applications. This enabled Temco to extract from other people's experiences the key issues of concurrent engineering relevant to the company itself.

The concurrent engineering implementation framework was underpinned by a series of assumptions:

- the complexity of concurrent engineering is determined by NPD complexity;
- firms differ in respect of their concurrent engineering 'need';
- firms have different starting points for concurrent engineering;
- the success of concurrent engineering is related to planning and implementation.

Complexity

The approach taken to implement concurrent engineering was based on an initial understanding of complexity in terms of the product and the user interface. The idea that new products might be differentiated according to their level of complexity is not new, and was first developed in the work of Clark and Fujimoto (1991) in their seminal study of the world's auto industry. The main premise of the Clark and Fujimoto framework is that different combinations of internal and external complexity of new products give rise to different issues in managing the development process. As a basis for the framework, new products are plotted according to two dimensions:

Complexity of internal product structure (CIPS) This scale represents the complexity of a new product from a manufacturing point of view. This includes the number of distinct components that make up the product, the number and complexity of production steps, the number of interfaces involved in the development effort, the level of technological difficulty and the severity of trade-offs among different components involved in producing the product.

Complexity of product–user interface (CPUI) This scale represents the complexity of a new product from the point of view of the end-user. This includes both the number and specificity of performance criteria which new products must meet, and also includes the subtle versus the overt and well specified characteristics of new products. So, for example, the automobile rates very high on the CPUI dimension because car owners usually have difficulty in expressing their needs directly.

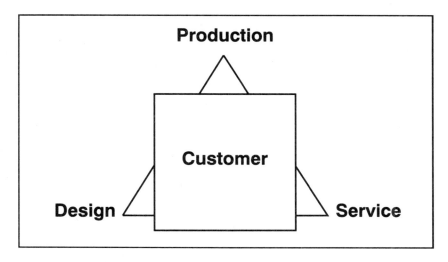

Figure 5.3 **Concurrent engineering integration (after Symmetrix)**

The relationship of these two factors is shown diagrammatically in Figure 5.3.

Although this framework is useful in drawing distinctions between firms based on their NPD complexity, the tool has not been used outside the original research. The application at Temco was the first attempt to use this in a practical format. The matrix was developed to allow a company's (or its product's) complexity relationship to be fixed. Firms that rate high on the CIPS and CPUI develop very complex products, whilst firms that rate low on the CIPS and CPUI develop the least complex products of all. It follows that the varying complexity of new products is reflected in the actual management of the NPD process. In other words, firms that develop relatively simple products have a relatively 'simple' development process, whereas firms that develop complex products have a 'complex' development process. The practical application of this relationship is realized by dividing the matrix into four quadrants, as shown in Figure 5.4.

Quadrant a Firms rate low on both the CIPS and CPUI and new products are relatively simple and comprise few subsystems and components. The number of processes used in the manufacture of the product is also low. Typical new products might include

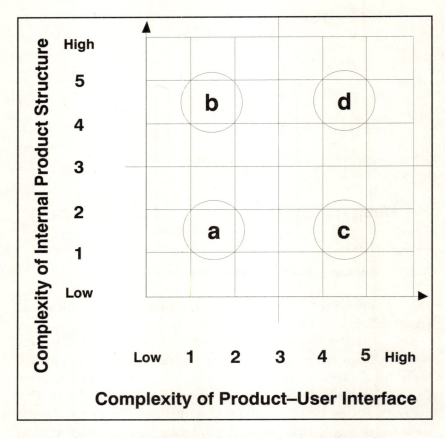

Figure 5.4   Concurrent engineering complexity assessment and evaluation matrix

household lightbulbs, twist drill bits, and general purpose nuts and bolts. In keeping with the relative simplicity of new products, the structures and processes that support the NPD process are also relatively simple.

Quadrant b   Products are very complex from the point of view of both the number of components involved in making up the product, and the number of processes used in their manufacture. Typical new products could include aircraft engines, dedicated machine tools, and automotive gearboxes. Quadrant b firms develop products according to customer specifications.

Quadrant c   Whilst products developed by quadrant c firms comprise relatively few subsystems and components, the difficulty lies in translating customer needs into attractive product concepts. Customer needs are sometimes latent and unexpressed, so it is important to monitor new developments in the market, and periodically to solicit the views of new and potential customers regarding new design ideas. Typical new products include consumer white goods and some forms of leisure goods, such as some keep-fit equipment and bicycles.

Quadrant d Firms in this quadrant develop the most complex products of all. This is reflected in the number of systems and subsystems that comprise the product. As a result, there are usually high numbers of staff involved in projects, and there is a crucial need for good coordination and communication. Products are also complex from the point of view of translating customer needs into design ideas. Typical new products include commercial aircraft and automobiles.

The framework

It can be seen that concurrent engineering affects all aspects of the NPD process, from the earliest stages of concept definition through to the eventual demise of new products. As such, a useful way of visualizing concurrent engineering is to conceive of it as a 'holistic' process which comprises a series of interlaced and interconnecting parts, consisting of people, processes, computer-based support tools, and formal methods. A depiction of the components of concurrent engineering is shown in Figure 5.5.

Thus the company's matrix placement and its NPD strategy and purpose determine the degree of need of each component part. For example, quadrant d companies are far more likely to use teams and tools and techniques such as QFD than quadrant a companies. In this way, the proposed concurrent engineering environment may be 'tailored' to suit the company's needs.

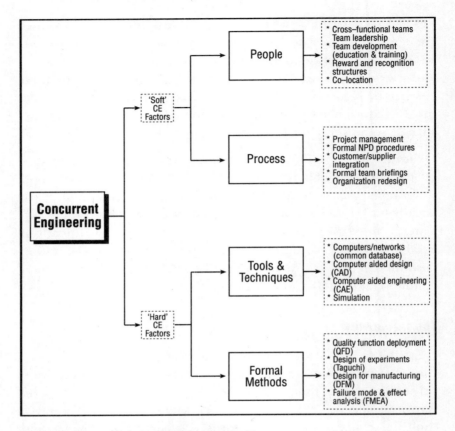

Figure 5.5 **The components of concurrent engineering**

The implemented concurrent engineering environment

From the complexity assessment and evaluation matrix, Temco's product development of the new-alloy based plated strand came out as borderline quadrant a or quadrant b, with a relatively low CPUI. The major CIPS determinants were the need to meet both the strict civil aviation requirements and the removal of the environmental problems associated with the old alloy. Secondary factors were the manufacturing characteristics of the new product and potential new uses. The evaluation showed that there was a simplicity of product coupled with extremely stringent performance requirements.

The next stage was a need to develop a framework that allowed Temco to 'tailor' its concurrent engineering introduction programme to produce an NPD environment that exactly suited its specific needs and circumstances. To do this, the methodology had to be capable of evaluating the company's concurrent engineering need and then allow a planned programme of development to be devised and implemented.

One of Temco's business strategies is growth through the twin processes of internal development and the formation of strategic alliances. The development of the new alloy-based product encompassed both these strategies. A major influence here was the fact that product development *per se* became a strategic issue, being formally considered on the strategic agenda of the company. It was not only the particular product (the alloy) that became strategically important, but also the actual development process itself. This emphasis ensured that there was senior management support throughout the whole concurrent engineering implementation process.

Being a small company, Temco's management style is informal and this culture is actively promoted as a means of involving and developing all staff. However, as part of a large corporation, its management systems are sophisticated, with an especially good management information system. By taking these factors into account, a form of concurrent engineering was devised that suited both the product's development and manufacture and the prevailing management processes and needs. The key elements of the concurrent engineering environment that was established at Temco are described by reference to a component model, shown in Figure 5.5.

People

The main problem with the 'old' NPD process was that it was functionally based, with responsibility for any development being handed sequentially from one function to another. This resulted in the typical set of problems associated with poor communications within the product development process.

Whilst all the participants in the development process

appreciated the importance of developing new products, the importance was always qualified from their particular functional responsibility. What was needed was a mechanism that facilitated programme management throughout the whole NPD process and, at the same time, was capable of being flexible enough to adjust quickly to the company's business needs. Many options were considered to resolve this problem, the major ones being:

- refine and improve the current system
- formation of an NPD team
- introduction of an NPD matrix programme management approach
- appointment of an NPD manager.

Refining and improving the current system was quickly rejected as an option. The system was not working and could not be improved sufficiently. It was also felt that a clear message as to the importance of NPD had to be sent to the rest of the organization. The formation of an NPD team was considered but ultimately rejected on two counts. First, the amount of development work probably did not warrant a full-time team. Secondly, Temco did not have the resources to support a permanent team structure.

Once the full-time team approach had been rejected, the next obvious possibility was the introduction of an NPD matrix management approach. This was rejected, partly for similar reasons to the full-time team argument, but more importantly because the fourth option was considered to be the one with the most potential. The final choice was the appointment of an NPD manager to have direct line responsibility to the managing director with a matrix-based team of other senior functional managers providing direct support to the NPD initiatives. The NPD manager would have the full responsibility for ensuring that a particular programme is agreed, fully developed and completed, whilst the functional managers would provide resource.

As with many of the best organizational development changes, this decision was partly taken from an appreciation of organizational need and an analysis of current organizational potential. The appointment of an NPD manager met all the essential requirements:

- Product developments would be focused.
- Responsibility would be clearly defined and accountable.
- All aspects of the total development process could be considered.
- Ownership would drive the process and thus shorten time to market.
- Senior management could monitor and control the process.

Process

There are several issues which occur under the process heading and they are particularly related to the aspects of concurrent engineering highlighted in this case study.

Supplier relationships

An important objective for Temco was the integration of suppliers and customers into the NPD process through the formation of strategic alliances. The company's activities are based on processing existing materials, resulting in limited R&D facilities within the organization. To develop a new alloy successfully, Temco need the assistance of a metal manufacturer with metallurgical expertise. The partner chosen for this work, a copper alloy producer, had a very good reputation for developing and supplying high quality copper and copper alloys, and had been a supplier to Temco for many years. It had an excellent development laboratory and a high metallurgical expertise coupled with advanced production facilities. Its product range also complemented Temco's specialist product range, for sizes and for materials.

Initially, the relationship was a traditional supplier/customer relationship, with formal communications. Throughout the development work, the relationship was refined through regular meetings, the agreement on common interests and strategic objectives and the development of compatible systems so that the material supplier became integrated into the Temco product development process. Throughout the work, the commitment of both parties to the project was strengthened, a situation only achievable through good communications and the realization of mutual benefit. The material supplier needed Temco to develop this alloy for its markets, just as Temco needed the supplier to provide the alloy and technical expertise.

The relationship has now been defined in law, through an agreement covering confidentiality and the approach to technical and commercial work. This agreement has ratified a relationship based on trust, through legal definition. Each of the parties has made their expectations and plans known and the relationship is strengthened through the sharing of aims and values. The two companies communicate at three levels: strategic decision making, technical development and the existing supplier/customer relationship.

Customer links

The stringent nature of the aviation industry's requirements introduced the importance of developing a product with the help of a 'safe' customer. This customer relationship was as important as the supplier relationship in the development of the new product. The new material was to be substituted in a very wide product range. It had many of the same characteristics as the original, but in some areas there were differences which may have affected the performance of some of the product range. To understand the new product fully, there had to be a dialogue between Temco and its customers.

In many cases, Temco's customers are not finished product manufacturers. The Temco product is processed further to make a component, which may then be used to make the finished product, or assembled as part of a system. This is the case for plated strand used for aerospace cable. Temco actively sought liaison with several of its direct customers and the ultimate end-users. This was done via direct contact and discussion, surveys of end-user requirements and the formation of an active group that encouraged dialogue and problem-solving opportunities.

New product development

Temco's NPD process was devised to suit the product simplicity and informal structure. For a small company, with the advantages of fast response and flexibility, care had to be taken in the design and use of such a system. Too formalized a system could restrain development and hinder the entrepreneurial management style. Each product development now follows the path described in Figure 5.6, with decisions based on market and technological

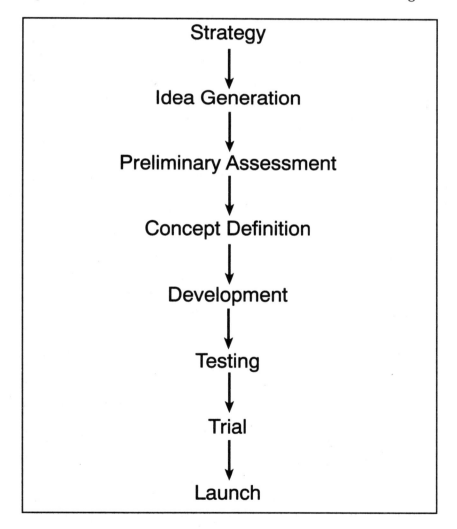

Figure 5.6　**Temco's concurrent engineering process**

information. With an informal management style, decisions may be made without reference to any structured aid. However, the general structure is used, and decisions recorded and acted on.

The stages of the NPD process are as follows.

Company strategy This is a prerequisite for successful new product development. The core competences analysis carried out prior to the plated strand development was central to the strategy development. Naturally, the subsequent product development activities were driven by the company strategy.

Idea generation Any new product concepts are evaluated against strategic and company-specific considerations, forcing Temco into consideration of all aspects of the product's development. At this point, the only financial commitment is to the continuation of development to the next stage of the process.

Preliminary assessment This covers both technical and market matters, and is used to determine the level of development work needed. This assessment is prepared by the NPD manager for approval by the supporting management team. The result of this stage is either the termination of the project or the transfer to the next stage.

Concept This is mainly the responsibility of the NPD manager. Here the project is defined and planned in detail and a budget assigned. The product is specified, and raw material and process requirements stated. Alongside budgets, milestones and timescales are allocated to be used as measures of performance throughout the development process. The potential market and product fit are investigated. Costs and prices are evaluated, and the major financial commitments of the project are made. The milestones follow the programme guidelines of risk minimization and incremental commitment of funds. They are used as decision points within the development process and agreed by the senior managers.

Development This is carried out according to the plan developed at the concept stage and is coordinated by the NPD manager. At each milestone, or at smaller intervals, the development is reviewed against performance parameters. Alongside the technical development, the market development plan is formed and executed.

Testing The first testing is done in-house, to known specification and performance parameters. When complete confidence is felt with the product performance, then the first customers can be approached. This must be a 'safe' customer, open to change and innovation. Once this stage has been reached, then the ongoing development work will travel at the pace set by the customer. The customer may have to go through stringent validation procedures, as dictated by the application for their products. Therefore, once the new product testing and adoption process has been started, it may take some time for the customer to proceed.

Trial Trial is defined as production volume runs. The product is still pre-commercial, and trials are used to smooth any problems and the NPD manager is directly responsible here. The reduction of risk coupled with fast time to market are still the major issues.

Launch This is the province of the marketing department and is planned in detail to ensure that the product has the best possible chance of successful entry to the market. The approach will differ according to the products and target market. Thus the approach is tailored to fit current business and marketing strategy.

Tools and techniques

Temco already used fairly sophisticated computer systems and techniques and little development was required in this area. However, a formal technique of identifying core products as the link between core competences and end products was implemented. This concept was applied to Temco to assess several skills and areas of expertise, through the following questions:

- Does it make a significant contribution to the perceived benefits of the end product?

- Does it provide potential access to many markets?
- Is it difficult for competitors to imitate?
- Is it essential to remaining competitive?
- Would losing this affect future business significantly?

Temco's core competences were found to be:

Material performance knowledge   This core competence is the expertise held by the people employed at Temco and it has been gathered over the period that Temco has been in existence, growing with the addition of new skills. The right people must be recruited and then kept within the organization. One way of retaining key personnel is by developing them alongside the company.

Stranding   The basic process of stranding can be used by most wire companies, but Temco has made advances over the base technology to gain advantage. To compete in high added-value markets, such as aerospace cabling, the strand produced by Temco must be of exceptional quality.

Plating (silver, nickel, tin and solder)   In the process of optimizing plating production, expertise has evolved. As a high added-value process, Temco has come to rely on these areas of expertise.

The concurrent engineering environment now recognizes and addresses these core competences and all future business development at Temco will be based on them to meet strategic growth objectives, through high added-value markets with high performance alloys.

Formal methods

There are basic performance requirements for aircraft cabling for which precise metrics can be defined. These include stability of performance at various temperatures, conductivity, weight, diameter, termination integrity, chemical inertness and break strength. Because of the exacting requirements of the aircraft industry, Temco has a highly developed portfolio of formal methods related mainly to quality assurance and testing. Thus little development in this area was needed for the new concurrent engineering environment.

However, as a new NPD process was being introduced, various other formal methods were evaluated. The most obvious one was Taguchi's statistical techniques, but these were found to be inappropriate due to the clearly defined product and testing requirements and product simplicity. Quality function deployment was tried but found to offer little additional advantage. This is mainly because the coordinating systems Temco developed have brought it very close to its customers and their needs are well known and defined.

Two recognizable formal concurrent engineering methods that were adopted were design for manufacture (DFM) and failure mode and effect analysis (FMEA). The DFM system adopted was relatively simple as all products go through the same basic process. In a similar way, the form of FMEA adopted was tailored to meet the basic product simplicity.

A review of the concurrent engineering implementation process

The aim of the work with Temco had been to design and implement a concurrent engineering environment suited to the company's business needs, structure and culture. A ten-step implementation module was applied as a means of ensuring that all the major contributing factors and influences were considered. This helped to ensure that the concurrent engineering environment ultimately established would meet the company's needs. This framework is shown in Table 5.2, with each stage described below.

Stage 1: Strategy Typical business growth occurs through many stages of managed development, each being an S curve linked by transitions. These curves move the business through the overall growth curve, as shown in Figure 5.7.

The earlier parts of business growth are achieved by internal factors, and are classified as organic, and the latter parts are

Stage	Description
1	Strategic consideration of competitive advantage.
2	Need for change identified with key improvement requirements.
3	Concurrent engineering assessed as being a reasonable improvement vehicle.
4	Senior managers understand concurrent engineering concept, process and related changes.
5	Evaluation of need, focus and emphasis.
6	Plans for concurrent engineering establishment, including improvement measures.
7	'Capture' of current process activities.
8	Implementation of plans (pilot?) and of process changes.
9	Evaluate results, especially via improvement measures.
10	Document results.

Table 5.2   **Ten-step implementation model**

achieved by acquisition. The three most important success factors for business growth, as defined by Page and Jones (1989), are investment in product innovation, hard selling and high profile leadership. Temco is currently focused on developing the business around the activities of product innovation, expanding distribution and investment in technology. The improvement in the capability of the company to innovate new products became a major strategic goal; hence the move to concurrent engineering.

Stage 2: Need for change identified   The stimulus to activity was directly related to the classic concurrent engineering change drivers of reducing time to market of new products, right-first-time design and integration of activities (internal and external). These needs were clearly identified by Temco's managing director.

Stage 3: Concurrent engineering as an improvement vehicle? Identifying the need to change is one thing, taking action is another. Companies often seem to be looking for the 'universal panacea' that will solve all their problems. However, a review of available improvement options is preferable so that the optimum route can be selected. This requires knowledge within the company which in turn necessitates external links. The

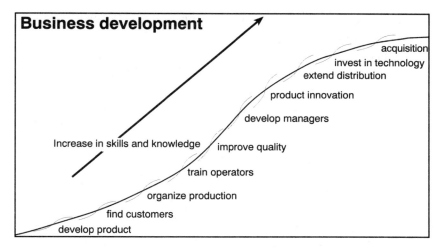

Figure 5.7 **Business development and growth curve**

collaboration between the company and the university authors was central to providing this information transfer.

Stage 4: Senior managers 'on board' The main person influencing the direction of product development was Temco's managing director who had already taken the decision that things had to change. Considerable time was spent in discussion between the managing director and the academic collaborators of the company's needs and explaining options and their implications. Ultimately, this led to the decisions and actions described in this chapter.

Stage 5: Evaluation of need, focus and emphasis A series of targets is required within a company to ensure all activity is focused on common goals. The evaluation of core competences carried out by Temco was crucial in determining the requirements and focus for change, both within the company and in relation to its external relationships.

Stage 6: Concurrent engineering establishment planning This is the critical stage of concurrent engineering implementation and required not only detailed planning of the form of concurrent engineering and its introduction, but 'selling' of the concept to company personnel. Much of the work on assessment and evaluation had been done and the form of the concurrent engineering environment and its major components were known, as were the likely impacts of the changes. Because the concurrent engineering form had been clearly assessed, it was relatively straightforward to detail the work that had to be done within the allowed timeframe. The overall implementation process was managed by the NPD manager and the managing director using a formal, monthly reporting system.

Stage 7: Capture of the current process   This stage is necessary to ensure that the proposed form of concurrent engineering does not miss any current activities. It is done at this stage so that the assessment and evaluation exercise may be conducted from a basic needs viewpoint (i.e. without preconceptions and constraints). In Temco's case, this stage was fairly easy to complete as the process was already well documented in its own right within the company's existing quality systems.

Stage 8: Implementation   It is reassuring to say that the reality matched the theory almost exactly and no significant problems arose during implementation at Temco. Although the new alloy-based product was a development in its own right, in essence it acted as a pilot programme for the concurrent engineering establishment process. The attention given to understanding concurrent engineering, explaining the concept to the organization and planning the implementation process all paid dividends. The consistent support from top management (especially the managing director) and the presence of a product champion (the NPD manager) led to a smooth change from the old to the new process. The new product was developed and marketed with significant reduction in development lead-time. There is no doubt that the effort put into the pre-implementation stages made the introduction of the new process go smoothly.

Stage 9: Evaluate the results   The programme has been an outstanding success. Not only were the product performance (technical requirements, manufacturing acceptability, toxicity etc.) and development timescales met, the strategic linkages with suppliers (especially) and customers are supporting the environment and acting to help maintain the momentum.

Stage 10: Document the results This is often a stage that is omitted but it is vital as a learning process and for future reference. A significant stimulus for this activity was the close collaboration between company and university.

Conclusions

The development of a new, high performance alloy for aerospace cabling was the strategic focus around which the new concurrent engineering environment was implemented. It arose from identified business needs and included collaboration with partners, customers and a supplier. The partners brought a knowledge of complementary technologies and markets to the relationship, whilst the supplier provided materials information essential to the success of the project. Indeed, without the input from the supplier Temco's competitive position in a core product area would have been weakened.

This was the first time that Temco had addressed new product development from a systems approach. In the past, technical and market developments have been isolated from each other, leading to lengthy projects with little communication within the business organization. The adoption of an integrated approach to product development proved to be an effective and rapid way of meeting new market needs where the 'time to market' factor has become essential for survival and growth.

The collaboration between the company and academe was central to the development of an implementation suited to Temco's particular circumstances. Using a centre of knowledge such as the university ensured that experience from many and diverse sources could be evaluated. This is in contrast to the situation where only internal knowledge is applied, which can result in a limited set of options.

As a consequence of this new approach, the development of new products has become more visible in the company and is no longer hidden within the technical or sales departments. The design and implementation of this concurrent engineering process has influenced the company culture as the need to successfully manage new product development has been recognized. Most importantly, product development within Temco is market led and integrated within the business development strategy.

The future

The success of the programme has been enhanced by its clear visibility within and importance to Temco. The NPD process put in place is fully supported by all the functions because it worked, removing problems that had existed for some time in this area of the company's activities. The complete process has been documented in terms of both application and record, and this is helping to ensure that the momentum is being maintained. Of particular importance here is the incorporation of the NPD process into the senior management monitoring and control system.

The success is also being exploited in publicity material for customers and within the BICC group, thus increasing commitment to the new process. This publicity has also had the effect of alerting other operational factories as to the potential benefits that might be achieved from using this approach.

Bibliography

Clark, K.B. and Fujimoto, T. (1991) *Product Development Performance: Strategy, Organisation and Management in the World Auto Industry*, Boston, Mass.: Harvard Business School Press.

Garret, R.W. (1990) 'Eight Steps to Simultaneous Engineering', *Manufacturing Engineer*, November, 41–7.

Lorenz, C. (1993) 'Stepping Out in a New Direction', *Financial Times*, 24 May.

Page, A.S. and Jones, R.C. (1989) 'Business Growth: How to Achieve and Sustain It', *Leadership and Organisational Development Journal*, **10**, (2).

Symmetrix Inc (1991) 'A draft discussion paper for businesses in the process of transforming an organisation to integrated product development', The Concurrent Engineering Research Centre, 23 September, 2–22.

Ziemke, M.C. and Spann, M.S. (1991) 'Warning: Don't be half-hearted in your efforts to employ Concurrent Engineering', *Industrial Engineering,* February, 45–9.

6

Controlling the Product Development Process

Rolls-Royce and Morris Mechanical Handling Ltd

In this final chapter of comparative case studies two very different companies describe their experience. Both Rolls-Royce and Morris Mechanical Handling are contrasted here since they had similar requirements of reducing unit product costs and to some extent progressed in related ways. In the first case study Rolls-Royce describes how, whilst in the past they had successfully achieved reductions in time to market of their aerospace engines, their unit costs remained too high. With each engine requiring in the order of 30 000 components, many of which were procured from external sources, the cost structure of the product was heavily affected by the component suppliers' ability to provide cost-effective designs.

The situation for Morris Mechanical Handling, a manufacturer of container-handling cranes, differed from Rolls-Royce in that the cause of cost pressure focused attention on internal design activities. Global pressures on cost were resulting from growing capabilities around the world, with manufacturing facilities being located close to the final erection site. Bought-in components for container cranes are semi-standard, which means that reducing final costs was primarily dependent on improving Morris's in-house activities.

Clearly, the two companies have many significant differences in terms of size, technology requirements and the time taken to

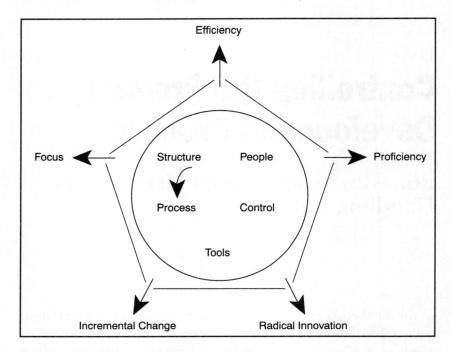

Figure 6.1 **The concurrent engineering framework**

develop new products. However, in terms of the concurrent engineering framework (see Figure 6.1) it can be recognized that both
companies were responding primarily to the pressure of Efficiency,
by tackling the Control element in the concurrent engineering
implementation. Whilst most if not all of the other pressures
illustrated in the concurrent engineering framework were
undoubtedly present to some greater or lesser degree, the focus
of activities was driven continuously by the need to reduce costs.

The solution

In response to the need to reduce cost, both companies put in
place measures which would allow them to gain improved control
through local responsibility. However, the approaches taken by
the two companies differed significantly and provide a very clear
contrast. In the case of Rolls-Royce attention was paid to
relationships with external suppliers as the way of reducing costs.

Morris Mechanical Handling, on the other hand, conducted a purely internal exercise to achieve the same ends.

The approach of Rolls-Royce was to initiate a programme of integrating suppliers more closely into the design process. Databases of supplier capabilities and knowledge were available to all Rolls-Royce designers so that they could consult with suppliers directly. At the concept design stage individual designers would communicate directly with approved suppliers to obtain estimates on cost and recommendations on design improvements. An iterative cycle would be established, with suppliers visiting Rolls-Royce as the component design was firmed up. By using this approach design options could be evaluated in terms of cost at the earliest stages of the design process. Designers who previously only considered functionality during the initial stages now saw cost information much earlier in the design process. The ultimate aim of reducing the total cost of the engine then became a reality under the control of the team of designers.

In contrast, Morris Mechanical Handling recognized that an internal initiative was more appropriate to its circumstances. The company developed local control of cost by establishing cellular operations at all levels within the company. Operational cells were supported by 'excellence' cells whose function was to implement change in areas of process and tools. Central to this concept were 'green areas' which provided immediate visual measures of individual cell performance. The displays of performance measures were not just concerned with internal cell activities, but additionally included cross-measures between cells. These cross-measures provided the required information to ensure that inter-cell activities could be measured, with action taken if improvements were necessary.

In both case studies the importance of changing the way people operate is highlighted. The new methods of measuring activity provided the control function, but it was recognized that for the changes to be successful it was necessary for them to be accepted by the designers. The two case studies illustrate alternative ways of achieving this end. In the case of Rolls-Royce it was through a process of training and education that the need for this new way of working was introduced. Long-term conventions were overturned through the demonstration of benefits to be achieved by early consideration of supplier input. In contrast, Morris Mechanical Handling demonstrates a route through change when the perceived security level is so high that change is very difficult to implement. An active policy of reducing security levels was implemented in order to generate the necessary impetus to achieve change. Once change was initiated then the level of security was increased in order that training initiatives could be introduced to a receptive workforce.

A final issue relates to the implementation methodology employed. In the case of Rolls-Royce the process of change followed the establishment of task-forces which were required to identify and optimize critical business processes. Although this was seen as a new initiative, entitled Project 2001, it built on expertise developed through previous business improvement initiatives. On the other hand, Morris Mechanical Handling embarked on a radically new activity with which it had no previous familiarity. It employed the concept of a viable system model to identify the various operation cells and control functions necessary to create a satisfactory organizational structure. Based around this concept it reconfigured its entire operations so they were based on a cellular system.

Utilizing Suppliers' Expertise to Reduce Unit Costs

Rolls-Royce

Oliver Towers

Introduction

Since its formation in 1906 Rolls-Royce has been an engineering company for which new product development has been essential to its survival and prosperity. In its early years, new product development was led personally by Henry Royce, whose eye for detail all the way from initial design concept to manufacture was legendary.[1] Indeed, Henry Royce had a close involvement in all Rolls-Royce's major engine projects until the end of his life in 1933. By the time that Rolls-Royce had the jet engine in production considerable expansion had occurred, some 40 000 people being employed by 1959. Needless to say, the involvement of one person in the detail of design, development and manufacture of one product became impractical, and clearly defined roles and functions were developed to be responsible for more specialized activities. However, the difficulties of a highly functional organization in responding rapidly to changing customer needs in a competitive environment are well known.[2] Like all other mature companies, Rolls-Royce has had to learn ways to adapt despite its size and history.

Concurrent engineering has been a powerful tool in this endeavour. This case study begins by giving a contemporary overview of the Rolls-Royce Aerospace Group and its operating environment, prior to summarizing the history of concurrent engineering practices as developed up to about 1990. A more detailed account is then given of Project 2001, an initiative started in 1991 to move concurrent engineering a further step forward in its evolution within the Rolls-Royce Aerospace Group.

Rolls-Royce today

Rolls-Royce is involved in the design, development, manufacture and support of a wide range of power systems. Its main products (60 per cent of 1993 turnover) are jet engines for powering civil and military aircraft. This activity is handled by the Aerospace Group. The other main organization is the Industrial Power Group, whose products also include gas turbines (in this case for electricity supply, pumping and ship propulsion), as well as steam generators, switchgear, transformers, boilers for electricity supply and lifting gear. Total company turnover was £3.5 billion in 1993, 70 per cent of which was exported. Of the manpower of approximately 46 000 at the end of 1993, 26 000 were in the Aerospace Group.

The Aerospace Group's current product range includes the RB211 family of engines, which power a wide range of aircraft including the Lockheed L1011, Boeing 747, 757 and 767. Other products are the Tay engine, which powers the Gulfstream IV, Fokker 70 and 100, the V2500 engine (produced in an international partnership), which powers the Airbus A320 and A321 and McDonnel-Douglas MD-90, the Olympus engine, which powers Concorde, the RB199 engine, which powers the Tornado fighter bomber, and the Pegasus engine, which powers the unique Harrier 'Jump Jet'. Currently under development are the Trent family of engines for the Airbus A330 and Boeing 777 and the EJ200 engine (in a European partnership) for the Eurofighter 2000. In addition, a joint company with BMW, BMW-Rolls-Royce, is developing the BR700 family of engines. The number of engines delivered each year is measured in hundreds, for example Rolls-Royce delivered 413 engines for civil aircraft in 1993.

Despite the simplicity of the basic design concept, a modern jet engine is a sophisticated piece of machinery. Typically it will have 30 000 different parts, which themselves vary in complexity from a state-of-the-art high pressure turbine blade made from a cast single crystal superalloy, to numerous bolts and washers, which despite their seeming routineness nevertheless have an essential function to perform. The complexity of a modern jet engine means that new product development is expensive. For example, Rolls-Royce made an investment in research and development (R&D) of £253 million in 1993, a relatively high proportion of which was for the development of the new family of Trent engines.

The Rolls-Royce Aerospace Group organization, shown in Figure 6.2, has developed in recent years to focus on the needs of specific customer groups, to increase local accountability for business performance and thus to maintain the company's position in world markets. Project directors are profit-accountable for their market sectors and report in to the managing directors of Commercial Aero Engines Limited (CAEL) and Military Aero Engines Limited (MAEL). In addition, Aero Engines Services Limited (AESL) has been created to focus on the high competitive aftermarket for jet engine products and services. These three trading units are serviced by the operating units (engineering, manufacturing business and procurement), which each have a director on the Aerospace Group Board. New product development is led by the appropriate trading group, using resource from these operating units as well as obtaining support from external suppliers and customers. At its peak a major new engine project would have around 1000 people working on it in Rolls-Royce, and this excludes the essential contributions of customers and suppliers.

Major competitors to the Rolls-Royce Aerospace Group include GE Aircraft Engines and Pratt & Whitney in America and SNECMA in France. The market has always been competitive, but never more so than in recent years, where the combination of

194

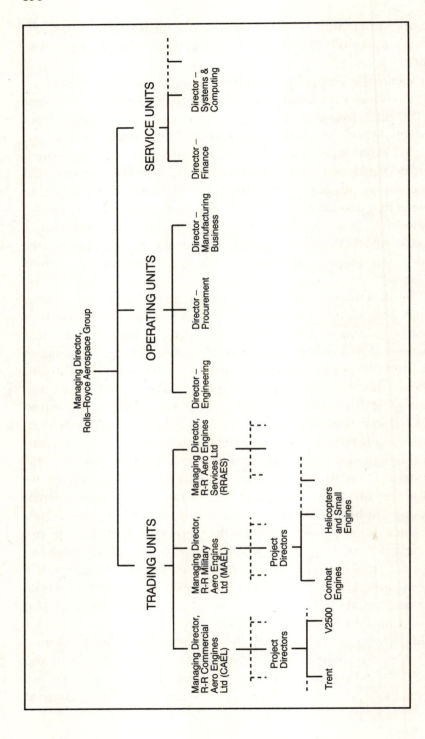

Figure 6.2 **Rolls-Royce Aerospace Group organization**

reduced demand in the military market and worldwide recession has left the competitors contending for a share of a reduced overall market. The increasing competition has made it more important than ever to develop new products rapidly whilst at the same time keeping the product cost to an absolute minimum.

Concurrent engineering in the Rolls-Royce Aerospace Group before 1990

The main drivers stimulating the early use of concurrent engineering were the need for improved ease of manufacture, reduced lead-times and reduced costs for product development, as illustrated in Figure 6.3. These are discussed in turn below.

Ease of manufacture

In the past the interface between manufacturing and engineering has been typified by the attitude 'If you can draw it, I can make it' within the manufacturing fraternity. This divide between the two activities encouraged the designers to design components with ever greater complexity and use more exotic materials to push their component designs to achieve the best possible performance and minimum weight. This increased the difficulty of manufacture, scrap rates increased and it became more

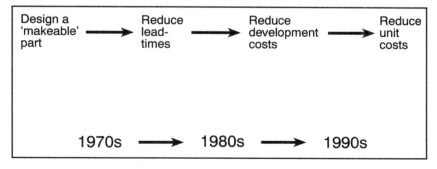

Figure 6.3   **Changing emphasis of concurrent engineering in Rolls-Royce aero-engine business**

difficult to achieve the timescales required for engine development. Increasingly the designers had to accept components to the practically achievable standards dictated by the constraints of the manufacturing processes ('product of process') and the corresponding limitations this placed on component performance.

It was recognized that the gulf between the design and manufacturing activities could not persist and initiatives were taken to improve the situation. First, a procedure was implemented which required a design to be 'bought off' either by internal manufacturing personnel or by manufacturing engineers within the purchasing function. Secondly, manufacturing engineering project managers were appointed and placed within the engine project teams, both to ensure that this buy-off procedure was understood and followed, and to act as facilitators to improve the interface between designers and manufacturing personnel. The third and most effective measure would be to create co-located teams including all the internal people involved in product design and development, i.e. designers, detailers, development engineers and manufacturing personnel. The practicality of this, and progress towards achieving it, is discussed below in the section entitled 'Co-location'.

With a clearly defined 'make or buy' policy it is easy to identify the relevant internal manufacturing personnel to involve in concurrent engineering. With external suppliers, however, commercial relationships need to be established and need to be on a firm footing before concurrent engineering can be at its most effective, and certainly the relationship has to be very close before co-location is practical. In some of the examples of concurrent engineering in the late 1980s, external suppliers were chosen very early in the project activity, and in some cases suppliers' representatives were located on Rolls-Royce sites, but not co-located with the design or project teams.

Lead-times

The traditional product definition process is 'end on end', i.e. when one person has finished their task they pass their output to the next person in the chain. The result of this, apart from the risk of designers producing unmakeable designs, was that the time taken to develop new components was prohibitive. A powerful tool for reducing these lead-times was to start performing some of the tasks in parallel, or 'concurrently'. Thus, for example, the manufacturing engineering evaluation and planning could start when the designer had proposed design concepts, rather than waiting for a detailed drawing. These work practices, since they rely so heavily on effective communications, were even more powerful when the different disciplines were co-located. Thus lead-times from initial design to approval of the finished component dramatically reduced. For example, this was a contributory factor in reducing the lead-time from five years to two and a quarter years for a high pressure turbine blade.[3]

Development costs

This case study largely concentrates on concurrent engineering between the design activity and manufacturing. The engineering effort to develop a new engine, however, calls on a wide variety of specialist skills, including aerodynamics, stress analysis and product testing. The increased use of computer simulations, to avoid expensive and time-consuming tests, and close interactions (i.e. teamwork) between these disciplines has been used effectively to reduce the cost of product development.[3] Increased emphasis is needed on the use of such tools to reduce unit costs in production.

Co-location

Teamwork and co-location have been referred to throughout this section. What has this involved in practice? The detailed answer is specific to individual projects and even individual components in the engine. General trends, however, can be observed.

One of the first trends described by Ruffles[3] was to integrate separate aerodynamics and cooling departments into a single Aero Thermal Technology Group. At a later date the mechanical technology personnel, involved in stress and vibration analysis and component 'lifing', were also organized on a component basis, thus creating 'centres of excellence' in turbine design and technology. These component-based groups (turbine engineering, compressor engineering, etc.) in some cases now comprise 100 people or more.

Open plan offices are widely used in Rolls-Royce Aerospace Group, for example 50 people from one of the above component groups are located in one open plan area. These groups focus on component-specific design and technology. Separately, engine project teams are created, with a mix of personnel from different functional disciplines. In recent years these teams have been located in special project halls, which can accommodate well over 100 people in one open plan area.

The component engineering teams' composition tends to be biased towards the design activity, whereas the project teams lean more towards engine testing and development. In both cases manufacturing and procurement personnel, as well as external suppliers, have tended to remain based at their local facilities (manufacturing and external suppliers) or in their commodity group (procurement), and to visit the co-located teams when required. The existence of the aforementioned manufacturing engineering project managers in the project teams is a step towards addressing this lack of a manufacturing expertise in the project teams, and indeed in one case a geographically remote manufacturing facility has a full-time representative co-located with a component engineering team. This tends to be the exception rather than the rule at present, and the remainder of this case study focuses on the design/manufacturing interface, since the design activity is when the decisions are made which determine the major costs of manufacturing.

Project 2001

In 1990 the Project 2000 initiative was launched in the Rolls-Royce Aerospace Group with the objective of eliminating non-value-added activities through the application of systems engineering techniques. The first step is to identify an 'opportunity owner' who can identify a major opportunity for improvement. The opportunity owner then establishes a support team, comprising senior managers who would ultimately be responsible for implementing any changes to work practices in their organization, and a task-force, comprising seconded personnel to work full time at improving the processes.

Project 2001 was the first task-force formed under the Project 2000 initiative. The original opportunity owner was the director of manufacturing engineering, who perceived a major opportunity to reduce the cost of manufacturing jet engine parts by making better use of suppliers' expertise. Her perceptions were that many of the engine parts were designed without recognizing the manufacturing cost implications of design decisions and that the company's ability to identify low cost manufacturing techniques needed improvement. A support team was established including senior management from the engineering, manufacturing and purchasing functions; a full-time, multidisciplined task-force was set up in March 1991.

The five-person task-force started by identifying the business processes underlying the perceived opportunity. These were essentially the processes for, first, supplier identification and selection and, secondly, technology identification and selection. Having identified the processes and proposed an ideal model, the task-force analysed the way these processes were performed in Rolls-Royce and elsewhere. These data were compared to the ideal model and used to redesign the processes to meet the overall objective, i.e. to reduce the cost of product manufacture. A cost–benefit analysis was performed prior to launching a pilot study to test and refine the process prior to implementation.

To get the effort and timescales in perspective, the phase up to the pilot study launch took eight months, and the pilot study itself took eight months. Why so long?

- It was the first task-force in the Project 2000 initiative and to some extent it developed and tested the methodology.
- This initiative crossed major organizational boundaries, potentially affecting many thousands of people in the workforce. 'Buy-off' of the proposals was essential, and

significant efforts were put into communication to achieve
this.

- The pilot was performed on parts being designed for a new
engine; the design activity occurred over the eight-month
period of the pilot.

The Project 2001 proposals

The process is schematically depicted in Figure 6.4. The
foreground process involves the design team working with
suppliers to develop the most cost-effective solution to meet the
required product performance, weight and cost requirements.
The background process involves the specification of future
requirements and the identification of technologies and suppliers
to meet these requirements.

Isn't this concurrent engineering? The answer is yes, but the
essential features of the process design which differ from previous
work practices are:

- the concept of a database of contacts at suppliers to which
designers can have direct access;

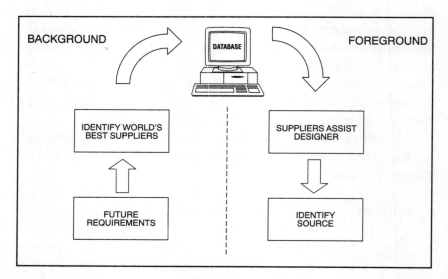

Figure 6.4 **Schematic depiction of Project 2001 proposals**

- the emphasis on minimizing unit production cost – previously the cost emphasis was on minimizing development expenditure;[3]
- the realization that a background process is required to identify potential technologies and suppliers and to develop these.

Without rapid access to experts at relevant suppliers, designers would never have time to consider alternative technologies. Indeed, they would revert to consulting their 'little blue books' of helpful contacts. The essential benefits of a corporate database are:

- it can ensure that suppliers have acceptable terms of business before they are entered onto the database;
- 'make or buy' policies can be enforced;
- designers can be preferentially directed to the best performing, commercially compatible suppliers.

To achieve all of the above means that preparatory work has to be done. Commercial negotiations need to have been completed. Technologies may need to be developed. Suppliers will have to be found. All of these necessitate the background activities in preparation for designing specific parts in the minimum possible time, to minimum cost, with the best performing, commercially compatible suppliers. In addition, the database of suppliers has to be created and maintained. This is not a minor task, with data needing to be sufficiently detailed to ensure that the appropriate suppliers are registered against the correct component types or manufacturing concepts, and with the data on supplier contact names, their telephone and fax numbers needing to be kept up to date.

Details of the concurrent engineering activities shown schematically in Figure 6.5 are described below.

Procurement strategy

Specifically this needs to address make or buy decisions. Which parts

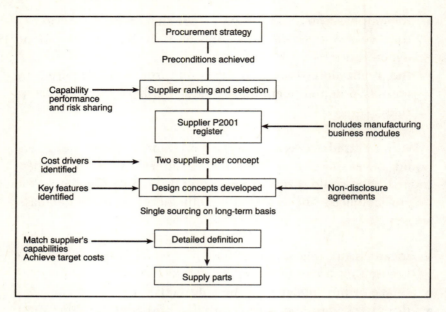

Figure 6.5   **Concurrent engineering to the Project 2001 work practices**

will be made in-house? Which parts will be made by external suppliers? Which parts will be competed for by internal and external suppliers? Typical factors which need to be considered in make or buy decisions include the competitive advantage gained from keeping technologies in-house, manufacturing capabilities, capacity and competitiveness (both in-house and at external suppliers) and the logistical requirements of an efficient supply chain.

Preconditions

Before considering suppliers as candidates it should be ensured that the companies are commercially compatible and that base quality control requirements can be met. Preconditions would include base contract terms, including those 'flowed down' from customers, as well as approval to the company's equivalent of ISO 9000.

Supplier ranking and selection

There may be many candidate suppliers meeting the preconditions. The next phase is to rank the candidates to select the two best performing suppliers to participate in concurrent engineering.

The Project 2001 proposals helped to stimulate the development and implementation of a supplier ranking and assessment system. This system measures suppliers in the key areas of delivery performance, quality, competitiveness, technology and management. The detailed factors considered are listed in Table 6.1. The results of the ranking are not only used for supplier selection, but also as a tool for identifying performance problem areas and developing action plans for performance improvement (the same system is used to rank manufacturing modules within Rolls-Royce as well as external suppliers).

Entry of suppliers onto the register

Suppliers who meet the preconditions described above and whose performance places them in the top two for a given technology are entered onto a database. The contents of the database are vetted by a senior management board on a six-monthly basis.

In due course the database will be directly accessible by designers as and when they are developing design concepts and need supplier input. At present, however, the designers ring an internal number to request supplier contact names. This 'help desk' is operated by the group within the procurement function which maintains the database.

- **Quality**
  Performance, responsiveness, administration
- **Delivery**
  Arrears, tailback, promise credibility, early delivery, responsiveness
- **Commercial**
  Cost reduction, competitiveness, risk sharing, administration
- **Technology**
  Process control, computing links, manufacturing strategy, capital investment, product capability
- **Management**
  Task, people, supplier management, delegated authority

Table 6.1 **Criteria for supplier ranking**

Concept development

A designer starts with a problem statement. This will cover component performance requirements, weight targets and a cost target. The designer will then propose a number of alternative concepts to solve the problem. They will then have to plan which concepts to pursue given the time available. The more time available, the greater the number of concepts which can be developed. To understand the cost implications of a given concept the designer needs feedback from prospective manufacturers. This is when the designer requests the supplier contact names. Having contacted the two suppliers for each design concept, the designer may ask the suppliers to visit to comment on the viability of the design, the cost of manufacture, both in terms of tooling (non-recurring costs) and repeated manufacture (recurring costs). When the designer has viable concepts, he or she will request price estimates from the suppliers. These will provide a breakdown of the constituent costs of manufacture, as well as an overall estimate of the total cost of manufacture.

All of these discussions take place with the mechanism of a non-disclosure agreement, signed by Rolls-Royce and the supplier, which prevents the disclosure of any data stated as being proprietary by either party.

With the aid of the data from the price estimates the designer can refine the alternate design concepts to minimize cost. Having done this the designer needs to evaluate the alternate design concepts and choose the optimum solution. This solution is drawn and documented to provide the basis for the two suppliers to quote. At the same time, the designer needs to establish the features which the suppliers would consider key to the component cost. These key features, when associated with the quote for the part, become contractually binding and are thus of commercial importance.

Sourcing

The two suppliers of the chosen design concept quote against an early part definition. These quotes are used as the basis of the sourcing decision. The winning supplier gains a long-term contract.

Detailed definition

The detailed definition of the component is produced with the aid of the supplier who will make it. The key features highlight to the detailer the design features which should be unchanged if possible. If the detailer does make changes affecting the key features the price for the part may need to be renegotiated.

Once the detailed definition is finalized the supplier makes the components and supplies them to meet the required delivery dates.

Lessons learnt on implementation of the new work practices

Some of the key lessons learnt during the implementation of the Project 2001 work practices are reviewed below.

Importance of testing

The systems engineering techniques used to develop the new work practices recommend testing by simulation. Where a new factory design has been proposed this can be done as a desk-top simulation, showing work flows through a physical model. The Project 2001 work practices, however, relate to complex interfaces between internal functions as well as with external suppliers. It was concluded that the work practices could only be tested through a live pilot study.

The pilot study focused on components already being designed over the same period. Eight components were chosen for the study. The supplier selection process was piloted, after a supplier seminar was held to explain the proposals to the candidate suppliers. The designers of the eight components were briefed on the work practices and continual dialogue was held with the designers to monitor progress and facilitate the process. The essential lessons learnt in the pilot were:

- the process needed 'fine tuning' to make it practical in a real operating environment;
- the designers, who lock in 80 per cent of the cost, needed encouragement to focus on unit production costs and help to understand the cost data;
- designers are unfamiliar with dealing with suppliers and the ethics involved;
- the process was fundamentally sound in delivering the benefits (i.e. reduced unit costs relative to past practices).

The essential point is that it would have been impractical to launch a full-scale implementation without the pilot because the practical problems involved in implementation would not have been realized and thus addressed.

Need for consistency and flexibility

The pilot, and subsequent experience with implementation of the new work practices, have demonstrated the need for people to modify and develop work practices to suit their specific needs. For example, the work practices can be operated independently of the question of who funds the tooling costs. The eventual contract, however, is quite different depending on whether Rolls-Royce's customer, Rolls-Royce or the supplier funds and owns the tooling. At the same time, however, it is important to identify the essential activities in the process which make it viable and to insist that these activities are performed in a disciplined and consistent manner. For example, all company employees are bound by the terms of the non-disclosure agreement. It is not discretionary.

These points are important. People will be much more receptive to new work practices if they feel that they are actively involved in championing and developing them. At the same time, however, they need to understand the base rules with which they have to comply.

Infrastructure

The Project 2001 work practices required various developments to be made to the infrastructure. Principal among these were the creation of appropriate procedures, the implementation of a supplier ranking system, and supporting data collection and analysis, a database for storing the Project 2001 suppliers and the creation and maintenance of a help desk to provide supplier contact names and answer general enquiries on the process.

Although the creation of these did not incur high levels of expenditure, they took time to coordinate and create. In addition, a commitment was required to ongoing costs to provide the help desk facility.

Communication and training

Communication is always identified as a critical factor in the success of any initiative. The implementation of the Project 2001 work practices has been no exception. Extensive briefings have been conducted across all the relevant functions, i.e. engineering, procurement and manufacturing, and supplier seminars have been held. Despite the maturity of the implementation phase, however, regular briefings on the project are still essential to increase understanding and spread the application of the work practices. Where briefings and awareness are insufficient to change people's motivation or develop their skills to operate the new work practices effectively, training is required. Considerable effort has been put into developing a training course which develops designers' awareness of the importance of unit cost and their influence on it, while training them in operating the new work practices. This two-day course has been delivered to 160 people.

Senior management commitment

Of all the requirements for successful implementation this is probably the most important. Without senior management commitment resources are not committed either to develop improved work practices or to implement them. Availability of resource, however, is not the only requirement. Leadership is also important.

Apparent commitment may well be visible at the outset of an initiative. It needs to be sustained throughout, particularly where the changes are significant in scope and particularly where major changes are required to the entrenched 'culture'. This requirement not only places a burden on senior management to show consistent commitment and support, but also imposes a responsibility on the implementation team to deliver the agreed plan and, equally importantly, to preserve the interest, understanding and belief of senior management in the initiative.

Applicability of experience to other industries

The introduction of concurrent engineering work practices into Rolls-Royce was driven by market needs: the need to reduce lead-times for product introduction, the need to continue to improve product performance and reliability, the need to develop a wider product range cost effectively, and the need to reduce product cost. Perhaps more than many companies, Rolls-Royce has in the past experienced the pain caused by delays in product introduction and the importance of prudent control of R&D expenditure! The financial crisis suffered by Rolls-Royce in the early 1970s undoubtedly increased the commitment to concurrent engineering as an important means of getting the required improvements in performance.

The complexity of a modern jet engine and the size of the team required to develop a new product makes it unfeasible for it to be done by one co-located, multidisciplinary team. Division of the task into logical groupings (either by specialist component technologies or by focus on major activities, e.g. engine test and development), together with the availability of large, open plan office accommodation, has enabled teams of in excess of 100 to be co-located. Larger teams may be feasible with appropriate accommodation, but the communication advantages start to reduce with such large open plan areas; subgroups are already present in the large project halls.

The limited involvement to date of manufacturing and procurement personnel and external suppliers in the co-located teams is an issue requiring further consideration. This remains a challenge because the internal manufacturing business, external suppliers and the procurement organization are focused on achieving world's best business performance levels on part families using similar groupings of manufacturing technologies. The internal engineering organization, however, is grouped by engine section (turbines, compressors etc.) which will tend to require many different part types.

Conclusions

Concurrent engineering has been used at Rolls-Royce for many years, but principally to reduce product development lead-times and product development costs. The competitive environment has increased the importance of driving down unit production costs. Concurrent engineering offers the framework to maximize use of the considerable leverage on unit cost at the design stage.

New work practices have been developed at Rolls-Royce to implement a concurrent engineering approach which is designed to minimize unit costs whilst achieving the other benefits normally gained, i.e. reduced lead-times and development costs.

Key factors to successful implementation are the use of a pilot to develop practical work practices from proposals, the creation of the infrastructure prior to implementation, comprehensive briefings, training and last, and most important, sustained and consistent senior management sponsorship and support.

References

1   Bastow, D. (1989) *Henry Royce – Mechanic,* Derby: Rolls-Royce Heritage Trust.
2   Womack, J.P., Jones, D.T. and Roos, D. (1990) *The Machine that Changed the World,* New York: Macmillan.

3 Ruffles, P.C. (1986) 'Reducing the Cost of Aero Engine Research and Development', *Aerospace*, **13**, (9), 10–19.

Local Control of Design in a Concurrent Engineering Environment

Morris Mechanical Handling Ltd

Bruce Norridge and Neil Burns

Summary

The processes that took place to change the Engineered Products Division of Morris Mechanical Handling Ltd are described here. The company restructured its design activities to improve business performance through the application of a participative approach to project control. In particular, this case study concentrates on the introduction of an innovative teamworking method known as the green area concept,[1] and some of the people-related issues and problems that had to be overcome for the successful introduction of this new system.

The green area concept is a disciplined system that is used by cell teams for the measurement and control of business performance. In Morris the complete business was transformed into a cell-based organization structure and the green areas were used by the members of each cell clearly to show their work performance, skills and improvements that were generated. The green area concept was extended by the company to include a complete business organization with a strategic green area and business improvement and development green areas. These various types of green areas were linked together into a communicating network. The design of the organization based around the green area concept was influenced by the viable system model first developed by Stafford Beer,[2] who proposed a five-system model for a 'viable' organization. The organization at the company was developed to ensure that the conditions for viability as defined by Beer were met as completely as possible.

Introduction

The Engineered Products Division of Morris Mechanical Handling designs and manufactures, to customer specifications, cranes for various applications up to a capacity of 250 tonnes. A large proportion of its business concerns container-handling cranes, but the company also supplies power stations, steelworks and other heavy duty applications. The division employs approximately 140 people at its East Midlands site in Loughborough, which it has occupied for over 100 years. Turnover in the division currently stands at approximately £30 million pounds per annum. The parent company has approximately 1000 employees, predominantly located in its manufacturing facilities in the UK and South Africa, but also has employees based in the Far East and Central America. The parent is itself a subsidiary of a much larger American crane manufacturing business.

The business is essentially an 'engineer to order' operation in that it designs and manufactures cranes to customer specification. Its ability to win orders depends on the quality of the relationship of the sales personnel with the customers, the reputation and history of past business performance, and the innovative features of the products. The initial stages of winning an order concern a competitive bidding process when tenders are prepared based on outline designs. If the bid is successful then a contract manager is appointed and the crane design and production process starts. The contract manager is responsible for ensuring that the crane is completed on time and within budget. It is commonly the case that contracts specify very high penalty costs for late delivery, since the installation of the crane is usually scheduled into major port upgrading activities. Late delivery of the crane will result in the complete project being delayed and significant loss of business for the port. Most contracts range in value from between £2 million and £12 million, with most of the larger ones, particularly port cranes, consisting of several identical cranes manufactured to the same design. The design and manufacturing activity from the time of the bid being awarded to the crane arriving on site typically runs from six months to one year for a single contract.

Assessing the need for change

On all contracts the primary object is to hit the time and cost targets, and therefore very considerable effort is spent in the early stages planning the complete project. However, in the recent past, the business had experienced very great difficulty in meeting these two main priorities and began to suffer accordingly. Due to increasing world competition and recessionary problems in the UK, Morris was being forced to take contracts with very tight performance targets and high penalty charges. Unfortunately, its ability to control these projects was proving to be inadequate, with the result that it was losing money on most contracts. A secondary consequence of the poor project control revealed itself in the form of high after-sales rectification costs. On-site activities became more costly during installation as a consequence of omissions occurring during the design and manufacturing activities.

The first action, and probably the most significant, that started the process of rapid restructuring was a change in the senior management of the business. A new managing director was appointed with the clear instruction to improve business performance. Early in the overall change programme the new managing director coordinated and implemented a strategic appraisal of all business activities based on a standard SWOT (strengths, weaknesses, opportunities and threats) analysis. This analysis provided the necessary knowledge on which to build a future vision for the business. Some of the elements of this were defined in formalized statements, such as:

> The company will be customer focused, able to meet customer requirements for an excellent product designed and manufactured within the time and cost specifications on the original tender document. As a company it will employ the best quality people in all domains of the business. There will be no 'them and us' and everybody will work together as a team ensuring that the customer is highly satisfied with the product and service and will regularly place repeat orders. Everybody will work for the excellence now and into the future of the business.

This vision was transformed into a number of statements defining objectives for change, which included:

1. To improve the time and cost control of the design and manufacturing processes.
2. To improve the skills of the employees and raise the professional standards of the business.
3. To introduce a new teamworking culture to ensure that everybody in the business feels part of the team and responsible for the business as a whole.
4. To encourage more innovation in new products or market approaches.

This in turn led to the determining of a series of actions defined for each activity domain in the business. The actions defined for the design activity were as follows:

1. To introduce operational design cells aimed at providing better local control of the design activities in terms of cost and time control.
2. To gain the full involvement of all the design personnel by changing and improving the organization culture to eliminate old 'them and us' divides.
3. To raise the skill base of the design activity by introducing a targeted training programme. The aim being to produce a

highly professional business with fully trained and competent people in all parts of the business.

4. To introduce modern technological systems to improve the design analysis and information flow.

5. To modularize the products, and the procedures used in designing the products.

6. To introduce risk funding for new product development that may produce a new customer response or new market.

7. To introduce a cell management scheme known as the green area concept. One of the aims of the green areas is to improve the local control of quality, cost and time whilst increasing overall personnel autonomy, and acting to initiate continuous improvement.

The change process in the design activity

To meet the vision of the future business it was considered important to improve the control of all the significant processes involved in the production of the product. To ensure the required level of control, each cellular unit involved in a contract would be linked into a supplier/customer network, and would each be responsible for the completion of a subset of activities.

Prior to the restructuring the design activity was primarily function and department based. Each function was built around the core design skills. For example, one department would contain mechanical design skills, another electrical engineering skills and another technical support, such as finite element analysis. The functional arrangement in the design office resulted in the work packages associated with a contract being moved from function to function. Each department would design its part of the contract almost independently. This method of working led to the usual problem of lack of communication and thus to design inefficiencies and errors.

Following a detailed appraisal of activities within each of the various design functions, a proposal was developed to introduce cell-based operations. Each cell would be responsible for a certain contract grouping. In most instances this grouping concerned product types, for example ship-to-shore cranes or steelworks cranes. However, groupings could be established based on geographical region if it was considered to be a major market where it was beneficial to demonstrate synergy between projects.

In each design cell there was to be a core cell team membership, consisting of a project manager and a senior mechanical engineer. The project manager had basic responsibility for crane production, which was effectively measured by the ability to meet full customer specification within the time and cost constraints defined in the original tender. The senior mechanical engineer was responsible for the technical aspects of the crane design and manufacture. The remaining cell team membership was dynamic in that it was dependent on the contract skill needs. These people were hired into the contract team from a technical support cell.

The green area system was used at this stage to enable the business to identify clearly the skill needs for the contract business over a period of time. This meant that the number and abilities of the technical support people were matched to skill need and to an averaged contract load. When, and if, the load became beyond that which could be covered by Morris personnel, then extra contract technical support would be hired into the company to meet the peak demand. This resulted in a cell team membership that consisted of permanent employees with core design skills plus temporary contract staff to meet particular needs.

The overall aim of the organizational change was to make the design activity more measurable and controllable, so that every part of the operation was streamlined, simplified, measured and locally controlled. A diagram of the new organizational structure, illustrating the relationship of the design cells to the other cells within the business, is shown in Figure 6.6.

Changing the culture

With such a massive change to the business structure it was obvious that there was going to be a need to make a major change

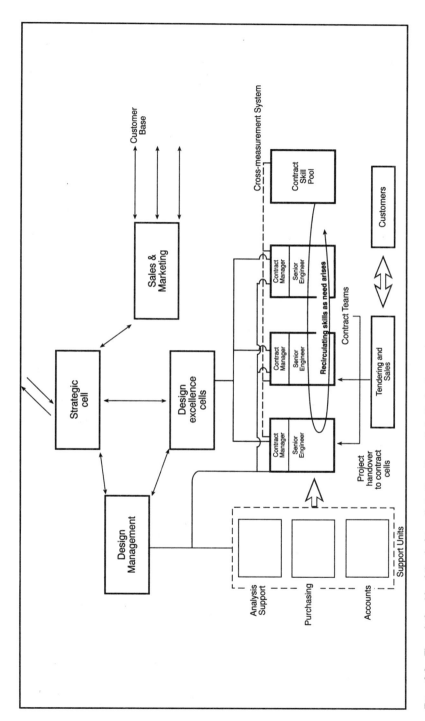

Figure 6.6 **The relationship of the design cells to the other cells in the business**

to working practices. In surveys of the design staff prior to change it was shown that, although they exhibited a high degree of interest in their work, considering themselves to be highly proficient, there was very little positive feedback between them and their managers. The style of management corresponded to 'command and control' with little opportunity for debate.

This directive style would be effective in obtaining rapid change, but it would not be appropriate for the more forward-looking, innovative and responsible workforce culture that was part of the vision of the future business. There had to be some cultural change within the business in addition to the proposed structural change. Increasing personnel involvement whilst at the same time improving the motivation of the designers was seen as central to the success of the programme. It was important that the culture of the organization was changed to one that would support the cell-based systems and at the same time meet the objective of continuous improvement. This implied change to the existing working practices and organization culture, as defined in Table 6.2.

When the new managing director arrived one of the first actions was to assess the situation in the design area. Once it was recognized that change was required, the next stage was how to progress it. Influential in shaping the assessment was the 'achieving change' diagram defined by Roger Plant in his book *Managing Change*.[3] An outline sketch of the diagram is shown in Figure 6.7. The concept is intuitively appealing. The implication

Old working methods	New working methods
Top-down control	High levels of discretion
Ready identification of failures but little recognition of good work	Managers as facilitators, not work providers and disciplinarians
Low levels of discretion	Everybody in the cell responsible for the performance of the cell and for generating local improvements
A culture of 'do what you are told to do'	Responsibility in the cell for training and personal performance

Table 6.2 **Changes to the existing working practices**

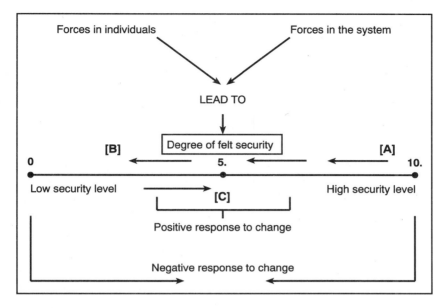

Figure 6.7 **Diagnosing readiness to change (after Plant)**

of the diagram is that the perceived security level of any employee is an important factor if change is to be readily adopted. If the perceived security level is low then change is unlikely. If it is high, then there will be too much complacency present for there to be any prospect of change. The perceived security level should be mid-range for successful change to take place.

The initial assessment, carried out by the managing director based on individual conversations with employees, concluded that perceived security levels in the design area were too high. The designers and their managers saw themselves as being secure and they did not feel responsible for the relatively poor performance of the business. In addition, it was recognized that there had been little change for many years in the design office and that the engineering skill levels had progressively dropped, mainly due to many senior design staff retiring. There had been no structured recruitment programme to replace these lost skills nor a programme of retraining for existing staff.

Clearly, there was a need for the skill base to be improved and there had to be changes to the organization to facilitate a new working culture. Initially, and primarily to achieve rapid change, the leadership adopted a directive style that enhanced the 'command and control' culture. This directive style lowered the perceived security level in the employees, as shown in the change from the right-hand side to the left-hand side of Figure 6.7.

As changes within the business progressed people began to be located within the new cell-based organizational structure. As had been previously identified, this type of structure could only provide the required benefits if a new culture was created. It was recognized that the new, low level of perceived security would not be compatible with a participative organization. In consequence, the managing director gradually modified the dominating leadership style to one which was more supportive. This resulted in the perceived security level of employees reversing direction in Figure 6.7. This rightward move continued until it was decided that the perceived level of security was roughly in the centre of the diagram.

Designing the cell-based system

The cell-based system for the design area formed an integrated scheme linked directly with the cells in other parts of the business. The organization of the system, particularly in terms of the information and measurement systems, was influenced by the viable system model (VSM) derived by Stafford Beer.[2] For Beer a system is viable if it is capable of responding to environmental changes, even if those changes could not have been foreseen at the time that the system was designed. In order to become, or remain, viable a system has to achieve a requisite variety of response to match the complex environment with which it is faced. The model has been in the public domain for a considerable time and it has been applied in many case studies. It was seen by the managing director as encompassing all of the ingredients for effective and viable operation of the business.

According to the model, for an organization to be viable it should have the best possible view of the environment relevant to its purposes. It should also have an organization structure and information system appropriate for that environment. The model is made up of five elements, or systems:

- System 1 – the operational activity
- System 2 – the operational coordinating activity
- System 3 – the operational management control system
- System 3* – the rapid auditing channel enabling the System 3 operational managers quickly to determine the performance of the operational system
- System 4 – the external environment scanning subsystem
- System 5 – the strategic level.

Figure 6.8 shows a schematic of the organization structure of the business overlaid with the five system levels of the VSM five-system model. This demonstrates that all of the requirements of the VSM were being met by the new organization.

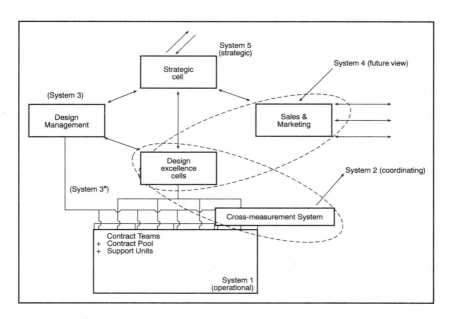

Figure 6.8   **Organization structure showing the five system levels of the VSM**

VSM System 1

The green areas in the design cells, the technical support cells and the administrative support cells, such as purchasing and accounts, formed System 1 of the VSM. Each System 1 green area was a self-contained unit with local measures directly appropriate to its work and supported by other displays which included such measures as attendance and skills that were present in the cell team. These simplified displays enabled the System 3* of the VSM, the rapid auditing of performance by the System 3 managers, to be carried out.

VSM System 2

Following the implementation of the design cell structure, another green area system was introduced as an overlay to the operational system. These new green area cells were known inside the business as 'excellence cells', a term adopted by the senior management of the organization. They were named 'excellence' because they were intended to make a major contribution to business improvement by implementing major changes and long-term developments. The membership of the excellence teams was made up from the operational green area team leaders, people directly drawn from the operational teams, and others from support teams with expertise appropriate to the improvement projects being considered at the particular time. Attached to each excellence cell was a facilitator or change agent who acted to implement some of the more complex improvement projects that required the participation of several operational cell teams. Typical projects would be the introduction of information technology systems in the business or design domain, or the introduction of advanced computer-based analysis tools. The excellence cell system acted as an integrating and coordinating layer around the System 1 operational cells and one of its roles was to ensure that ideas generated in one cell were passed to all the others, hence enabling an organizational learning role. The excellence cell team met on a fortnightly basis in order to achieve the coordination of business improvement.

In addition to the excellence cell team meetings, a second method of coordination was established. This consisted of all the team leaders of the operational cell teams getting together and holding a short review meeting at the end of each working day. The purpose of this was to raise any issues that applied across a number of the cells. This daily meeting ensured that day-to-day coordination was therefore achieved across cells. The combination of daily and fortnightly meetings ensured that the business operated with a unified purpose in meeting strategic objectives and therefore fulfilled the System 2 role in the VSM.

VSM System 3

The operational management of the design activity met on a weekly basis with all the cell team leaders to define operating procedures. This effectively established the System 3 level of the VSM. The workload for the operational cells was defined and any implementation requirements of policy from the strategic cell team passed on. The operational management had ready access to information about performance via the display boards in each of the operational green areas. The operational design management reported to the divisional manager who was part of the strategic cell team. This then formed the link between the System 3 function and the strategic System 5 function.

VSM System 3*

At the start of the working shift, in each operational cell there was a meeting to consider work problems, training and cell improvement issues. Part of the green area contract agreed by all the personnel in the business was that if any issue raised by a team member could not be answered by the team leader or other team members, then a notice was to be placed on the green area board. Senior managers (generally core members of the strategic team who regularly monitored the boards) would then be required to respond within 24 hours. The display boards in each green area, together with the requirement for a quick response to issues raised by operational cell team members, met the requirements of System 3* (the audit channel) in the VSM.

VSM System 4

The excellence cells formed the forward-looking System 4. They were primarily responsible for the improvement to the design activity itself and the products. To do this they had to have links with all the other systems and in addition to the outside world, particularly in terms of customer trends and competitor actions. Based on this knowledge they were able to define the improvement projects needed in their domains. They would then make their case to the strategic cell team, which would then prioritize in relation to overall business strategy and availability of resource. The excellence cell team would then plan and coordinate implementation of projects with results being registered in the excellence green area.

VSM System 5

The purpose of the strategic cell team was to determine business direction and strategy and ensure that the defined set of objectives was implemented. The membership of the strategic team was the business leaders, representatives from sales and marketing and team leaders from the various excellence cells across the business. The strategic cell therefore met the System 5 requirement in the VSM.

Green area performance measurement

An important part of the new cell system in the design and engineering domain was the measurement of performance. A typical range of measures in a design and engineering cell would be as follows:

1. Contract performance in terms of cost and time control.
2. After-sales rectification costs.
3. Personnel attendance at the morning meeting in the cell.
4. Skills available and skills needed to complete the contract.

5. Cross-measures, where the suppliers and customers to the cell measure its performance.
6. Other measures at the discretion of the cell team: one example may relate to the number of improvements generated by the team.

In addition to the direct measures of performance, the cross-measurement system was an important part of the new design activity improvement system. It provided an impartial 'outsider looking in' view of the cell's performance. The quality of the service 'from and to' the cell was then monitored and controlled by the cross-measurement system. Some typical cross-measurement linkages are shown in Figure 6.9.

Green area displays

Although there was a standard requirement for certain measures that should be displayed, team members were given considerable

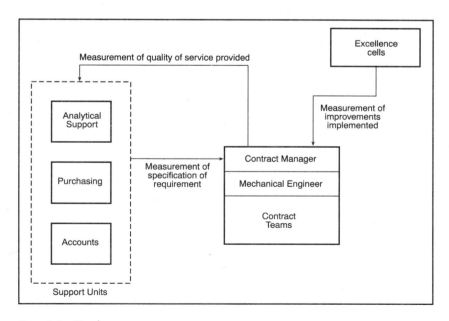

Figure 6.9   **The cross-measurement system**

discretion in designing the displays and setting up their own green areas. Each team decided on a name for its area, arranged its displays and kept them up to date. A diagram showing a typical green area display of measures is given in Figure 6.10. Every cell had several displays: an attendance register which logged the cell members' presence at a meeting at the start of the working day, a skills register which showed skills and training, relevant local and cross-measures, and improvements generated by the cells.

One of the most significant displays in the operational teams' green areas was that for skills and training. A multilevel scheme was developed to provide measures varying from the highest level of being fully professional in a particular skill, to the lowest level of being 'in training'. Each level was validated by a testing scheme administered by senior chartered engineers in the business.

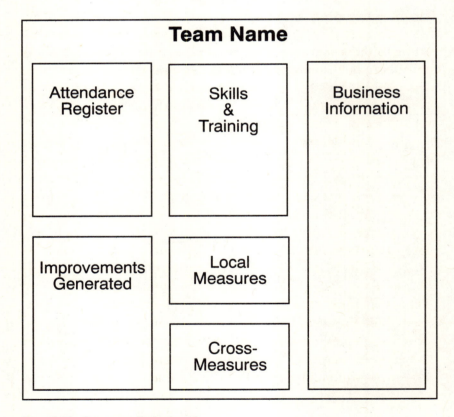

Figure 6.10   **Green area display format**

Specific professional skills, for example 'structural engineering', were then graded for each individual. At the start of each contract planning process the contract managers made an assessment of the skills needed to complete the contract and they matched this to the availability, initially within the cell and then within the business. This enabled them to define their optimum contract team structure and the desired cell membership. They could also define any training requirements needed to complete the contract.

The results achieved

The managing director initially adopted a directive style of leadership to obtain the major change to the green area system. There was relatively little cost involved in setting up the green areas and they were implemented very quickly, over a period of two weeks. However, the process of getting the areas utilized to full effect took considerably longer. Initially some champions were identified and these people often took a leading part in promoting the change in working practice that enabled the green area concept. Many of these initial champions went on to become team leaders. However, it took over a year before the areas were being used to full effect and the local measurement system was being used by everybody. The most difficult area was the cross-measurement system. People were reluctant to put formal measures on the display about the performance of other supplier customer cells. Although the number of these measures has gradually increased, especially in those cells involved in providing a service to other cells, there is still a long way to go with the cross-measurement system.

The structural change in the business organization to cell-based systems provided much better local control of performance and consequently enabled the products to be produced within time and cost estimate. This resulted in significant improvement in business profitability, as indicated in Table 6.3.

Contract Time and Cost Control Improvement		
	Averaged previous contract	**New contract**
Cost control of the contract	20% overrun each contract	10% overall reduction in estimate times
Labour costs	30% overrun	20% reduction in estimate times
Purchased parts price control	10% overrun due to unforeseen needs	10% reduction in estimated costs of purchased parts
Time for contract implementation	10% overrun	Easily met estimate times

Table 6.3 **Measured improvement in performance**

Lessons learnt

Resulting from the strategic analysis the business determined that time and cost control was of paramount importance and it changed its operating systems to reinforce this message.

The change in working practice and culture in the design office together with the improved business control were facilitated by the simple and very visual measurements of performance displayed in each green area.

The cell-based business resulted in much better control of operations and recent contracts have all resulted in significant profits being made. However, to obtain lasting change it is important that the leadership style is supportive of culture change. The new managing director amended the leadership style in response to changes in the perceived security levels of the employees. From an initial very directive style aimed at initiating rapid change, there was a shift to a more supportive role in order to obtain responsible delegation and continuous business improvement.

It is generally difficult to assess fully the success of any leadership style change because other factors are also always operating. For example, morale noticeably tends to go up when the business is perceived as being successful and winning orders.

However, it was clearly demonstrated within Morris Mechanical Handling that after the move to cellular working, and in particular with the introduction of the excellence team structure, a much higher rate of improvements was generated from the operational cell teams.

There was evidence of a much higher level of involvement of all the personnel in the design and improvement of their respective cell systems. There is now a much higher level of involvement in decision making by all employees at all levels. The results of this process of change indicate that important consideration must be given, simultaneously, to the human and technical redesign issues if lasting business improvement is to be achieved.

References

1   Middleton, B. (1990) 'The Nissan Green Area System', presentation to South African Society for Quality, Eastern Cape Branch, 7 August.
2   Beer, S. (1972) *Brain of the Firm*, London: John Wiley.
3   Plant, R. (1987) *Managing Change and Making it Stick*, London: Fontana.

7
Learning the Lessons

Chris Backhouse and Naomi Brookes

The case study companies

The original starting point for this book was the concept that concurrent engineering should not be considered as a uniform solution to improving the product introduction process. Different concurrent engineering solutions are needed by different companies to satisfy their own particular situation and set of requirements. In order to clarify this issue of differentiation, a framework has been used as an aid to highlighting the various key issues that should be addressed in any implementation of concurrent engineering. This framework relates the pressures acting on companies to change their product introduction system and the elements within the organization that go together to form a concurrent engineering solution. Whilst the framework does not provide direct answers to the problem of which form of concurrent engineering to implement, it does enable managers to structure their approach to implementation in a logical way.

The case study companies do indeed demonstrate a wide variety in terms of the form of concurrent engineering that they have adopted. Some companies have used a great deal of computer aided technology in their design and development of new products, whilst others have used very little. Some companies have implemented highly autonomous, full-time, co-located project teams, in contrast to those which have implemented lightweight teams consisting only of project managers with no other direct allocation of personnel. These differences, expressed in terms of Tools and Structure within the concurrent engineering framework, are matched by differences in the other elements of Process, Control and People. It is possible to highlight the key differences and similarities for all the case studies and this is shown in Table 7.1. In this table each of the five elements of a concurrent engineering solution is considered in turn for each company.

Concurrent engineering at Marconi

Marconi has adopted a highly autonomous business unit structure supported by core service departments. Within the business units a team structure operates to progress development projects. Team members are in the main co-located, comprising representatives from various design and engineering functions. Other team members are coopted from the support units to provide expertise in such areas as technical publications, manufacturing and sales. The use of highly sophisticated tools was not highlighted within the case study since there was a strong perception that the priority was to achieve a suitable product introduction process and then to move on to the issues related to support tools. The formal product introduction process, described in the case study, has the three main stages of investigation, development and release for manufacture. It is controlled by a series of formal milestones which are reviewed by both senior managers and peer groups. The implementation of concurrent engineering followed the path of demonstrating the benefits through a single, tightly focused project and then disseminating the approach throughout the business.

Concurrent engineering at Lucas Aerospace Actuation Division (LAAD)

This company has in some ways adopted a similar approach to concurrent engineering as that taken by Marconi Instruments. However, it has chosen to adopt an approach to structure that lies

	Rolls-Royce	Lucas Aerospace Actuation Division	Temco	Measurement Technology	Morris Mechanical Handling	Instron	Storage Subsystems Development, IBM	D2D	Marconi
Tools	High use of CAE, communication technologies & quality tools	High use of CAE, communication technologies & quality tools	Comparatively limited use of CAE & electronic communication significant use of quality tools	Comparatively limited use of CAE, quality tools & electronic communication	Comparatively limited use of CAE, quality tools & electronic communication	High use of CAE & communication tools	High use of CAE, communication technologies & quality tools	High use of CAE & communication technologies	Comparatively limited use or CAE, quality tools & communication
People	Roles largely unchanged with a high degree of specialization	Roles largely unchanged with a fair degree of specialization	Roles largely unchanged with some specialization	Roles of team leaders created	Roles largely unchanged	Roles unchanged with a fair degree of specialization	Roles largely unchanged with a fair degree of specialization	Roles largely unchanged with a fair degree of specialization	Roles largely unchanged with a fair degree of specialization
Structure	Very largely autonomous teams with wide membership	Highly autonomous team with wide membership	Lightweight team with wide membership	Autonomous team	Autonomous team with very narrow membership	Heavyweight team with wide membership	Heavyweight team with wide membership	Heavyweight team with wide membership	Autonomous team with wide membership
Process	Very formal process with a high degree of parallelization	Formal process with high degree of 'built-in' parallelization	Formal process	Formal process with some parallelization	Comparatively informal	Formal process with some parallelization	Formal process	Formal process with some degree of parallelization	Formal process with some degree of parallelization
Control	Milestone control by a formal review	Milestone control	Milestone control	Milestone control	Formal visual cross-cellular control	Formal review points seen as handovers	Panel review of milestones	Milestone control or projects	Review of milestones including peer review

Table 7.1   Concurrent engineering in the case study companies

further towards autonomy on the spectrum of teamwork styles than does Marconi. Many of the core service units which were maintained in Marconi's structure are permanently incorporated into LAAD's teams. The company's team structure consists of full-time co-located members who represent design, draughting, manufacturing engineering, stress performance and a commercial representative. Employees can be identified as having maintained the sort of skill mix that was in evidence before concurrent engineering was implemented, but their overall skills are now much broader. The type of product introduction process adopted is, as seen in the case of Marconi, very formal, but is deliberately specified to avoid end-on-end activities and to encourage a high degree of parallelization. LAAD also makes extensive use of CAE and communications technologies to support product development, but yet again takes a similar approach to Marconi in that it aims for process improvement before introducing new and advanced design tools.

Concurrent engineering at Instron

Instron decided that the most appropriate approach to concurrent engineering was to retain an identifiably functional organization with project teams being created across the functions. The company made a conscious decision to adopt this approach, rather than that of completely autonomous teams, because it felt that this better maintained the specialized skills of its workforce which were viewed as key to the company's success. The membership of teams is usually wide, to include mechanical, electrical, electronic and software engineers plus commercial functions. Parallel development of new products at the two main company sites in America and the UK is facilitated by communications technology and a set of product standards which define all component interface requirements. A formal process for product introduction is also in place within Instron which is controlled by a series of reviews defined as acceptance milestones. Despite the formality of these 'handover' points, significant parallelization of activities is encouraged, not least by the team structure itself. Instron makes significant use of computer aided engineering tools with high-speed electronic communication allowing the two development sites to collaborate closely during design optimization.

Concurrent engineering at D2D

The approach to concurrent engineering adopted by D2D has very strong parallels with the approach adopted by Storage Subsystems Development at IBM Havant – see below. D2D uses heavyweight teams superimposed on what is basically a functional structure at the local level. This approach is facilitated by the extensive use that D2D makes of electronic communication technologies to bring the various employees working on a project into a 'virtual team'. It is not uncommon for team members to meet each other physically only on rare occasions. For team members with limited input a physical meeting may never occur. As an electronics manufacturing service supplier D2D naturally makes significant use of CAE tools which allow for a high degree of design automation. It has a very formal product introduction process monitored through networked project management software.

Concurrent engineering at Storage Subsystems Development, IBM

Storage Subsystems Development has taken a very similar approach to product introduction to that taken by D2D. It is based on communications technology to facilitate rapid transmission of data, so that the physical location of team members becomes irrelevant in determining their level of input. This section of IBM operates a matrix management structure pulling together team members from the various functional groupings. The teams can comprise members from manufacturing, mechanical design, firmware, hardware and software design and test engineers. The process of developing new products is controlled by a series of reviews conducted by a panel formed for this specific purpose. As would be expected from a company such as IBM, Storage Subsystems Development makes very substantial use of CAE, computerized data management and electronic communications. It employs sophisticated configuration control systems and, as highlighted in the case study, aims to shorten development time by statistically identifying product confidence levels to minimize testing time. The company also makes considerable use of quality techniques such as Taguchi and quality function deployment.

Concurrent engineering at Measurement Technology (MTL)

Despite being a comparatively small company, MTL has adopted a very autonomous structure. It perceived the need to change the roles of people within the organization to create specific posts to promote concurrent engineering activities. The case study describes how the company developed a formal new product introduction process at the same time as several other key business processes. It was notable that it felt that the majority of benefits from implementing concurrent engineering could be achieved through the development of teams. A minimal number of new tools was actually introduced during the period of change, since the view was that the key factors concerned people issues within the company.

Concurrent engineering at Temco

Temco has adopted a very different approach from any other of the case study companies to the structure it has adopted for new product introduction. It used an extremely lightweight project team structure. This team has only one full-time member, the project manager, who is responsible for obtaining any other support that he or she requires from other functional heads within the organization. However, as with the majority of the case study companies, Temco has adopted a very formal new product introduction process, with control of this process achieved using milestone points. The adoption of new tools has not focused on high levels of computerization but on the introduction of quality tools such as QFD.

Concurrent engineering at Rolls-Royce

Rolls-Royce has adopted an approach to concurrent engineering that is characterized by the use of complex tools combined with autonomous organizational structures. The company uses full-time, autonomous, co-located teams whose membership comprises design and development engineers with manufacturing and purchasing representatives. These representatives act as a conduit to the appropriate internal or external manufacturers. The team is very large – often of the order of 100 people. The team is functionally supported by other specialists such as stress and aerothermal technologists elsewhere in the organization. No major changes were made to people's roles on the introduction of concurrent engineering and the skills managers with a particular role have remained substantially the same. The process adopted by Rolls-Royce to introduce its new products is highly formal and is controlled by a series of milestone events. However, the formality of the process does not preclude significant parallelization of activities in order to shorten product development times. As described in the case study, the concept of concurrent engineering includes close integration of suppliers within the product development process, with Rolls-Royce designers being directly responsible for obtaining product cost data from potential suppliers.

Concurrent engineering at Morris Mechanical Handling (MMH)

The MMH case study describes the introduction of a highly autonomous structure for its new product introduction based around the concept of 'green areas'. Team membership for MMH is relatively narrow, with only a few members within each operational cell. Operational cell membership usually comprises only project managers, mechanical engineers and electrical engineers. Other engineering services such as structural analysis form support cells. Manufacturing was considered as external to the system since many of Morris's products are fabricated close to the point of crane erection. Like Measurement Technology, MMH focused its introduction of concurrent engineering around the introduction of a new structure. It used this structure as a spur to introduce relevant controls focused around the 'green areas'. The roles that people adopted and the tools that they used also remained substantially similar in the operational cells.

It is interesting to note both the similarities and the differences in the way that the case study companies have implemented concurrent engineering. The major differences occur in the way that the companies have used tools and the type of structure that they have adopted. The use of communication tools has varied, from those companies which see IT as the basic facilitator for establishing teams to those who identify co-location as the prime requirement for good communications. In terms of CAE, it is clear that companies have employed the level of sophistication appropriate to their particular circumstances. In none of the companies could it be said that the CAE approach taken was beyond that which could sensibly be expected.

The types of teams implemented in the various companies has been seen to vary from the extremely lightweight to the very autonomous. Some companies have deliberately retained functional groupings to ensure skills retention. Other companies, usually those that develop larger teams or that have a limited range of products, can afford to move away from the functional groupings towards the autonomous team structure.

A major similarity between the case study companies can be seen to be the adoption of some type of formal new product introduction process which also includes a review of that process using a milestone technique. This is not surprising, since this approach has long been recognized as having a strong correlation with successful product development.

The concurrent engineering framework

In Chapter 2 a concurrent engineering framework was developed to help illustrate the relationship between external pressures felt by companies and the form of concurrent engineering they implemented. It was argued that the different ways in which these

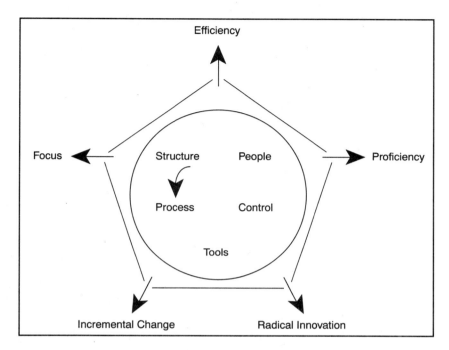

Figure 7.1 **The concurrent engineering framework**

pressures acted went a long way towards explaining the variety in concurrent engineering solutions. It is useful to consider each of the forces in turn and examine their presence in the case study companies.

Efficiency

It is hardly surprising that this force was found in all of the case study companies. After all, they were all case studies of concurrent engineering introduction and it is safe to assume that all the companies wanted the performance benefits that this would bring. What is interesting is that in different companies, different elements of efficiency were emphasized. Reduced time to market appeared to be a universal driver, but in some companies, such as Rolls-Royce and MMH, the primary objective was the reduction in unit costs. In other companies cost was not as heavily emphasized as the issue of maintaining or enhancing product quality. In general, it is usually the case that all companies experience cost, quality and time pressures on a regular cycle. Whilst in an ideal world these three elements should be attacked in parallel, reality does not always lead to this approach.

Focus

Again, the force of focus could be found in all of the companies but with obvious differences in the way that this was driven. In Rolls-Royce, Temco, MTL, LAAD and Marconi a very strong driver towards the adoption of concurrent engineering was in place. Each of these companies had strong executive support and a full-time team to introduce concurrent engineering by piloting new working practices on a real product introduction. Other companies tended to have a less direct force driving them towards concurrent engineering. In some cases the adoption of concurrent engineering practices was part of a wider initiative by senior executives to introduce teamwork. In other companies significant emotional events were in evidence, such as the appointment of a new chief executive or becoming a stand-alone company after being part of a larger organization.

Proficiency

The proficiency force is especially evident in companies which are involved in design-to-order activities. The most striking example of the need to demonstrate proficiency is seen in the D2D case study. This company must be able to demonstrate its capabilities as an electronics manufacturing services provider, doing this through its proven track record and infrastructure capabilities. These are highlighted in D2D's use of communication technologies and sophisticated database management. Morris Mechanical Handling, another design-to-order company, demonstrates its proficiency by its development of skills capabilities focused around its new team structure and illustrated in its green areas. Rolls-Royce demonstrates its proficiency via a proven track record in order to provide its customers with sufficient confidence to commence development projects which could run for many years.

Radical Innovation

Marconi, Lucas Aerospace Actuation Division and Rolls-Royce all felt the effects of this force when they saw the shrinking size of their traditional military markets. They had to react to radical changes in the make-up of their customer base and respond accordingly by significantly changing their products and their approach to development. In a similar way, MTL and Temco experienced changes in their market which necessitated radically new types of products. Temco had to develop new relationships with customers and suppliers to ensure the successful development of a new product based on novel materials. Measurement Technology experienced a rapid change in market pressures caused by a change in communications technologies, necessitating the development of new intrinsic safety equipment on a much reduced timescale.

Incremental Change

The force for incremental change is best represented in the case studies by the way that Instron, D2D and IBM have introduced new products. These have tended to be extensions of their existing products and have been introduced by systems that have changed in an evolutionary and not revolutionary fashion. This incremental approach is highly relevant in particular commercial environments, especially where a company can monopolize a previous radical innovation. When looking for the evidence of the force for incremental change, it is important to remember that the judgement on whether or not a change is incremental will be relative. An incremental change for a large, technologically complex company may be a radical change for a small, unsophisticated organization.

Concurrent engineering: what works where

The fundamental premise of this book has been that concurrent engineering solutions must be different according to the specific circumstances of the environment in which they are going to operate. The variety of forms that concurrent engineering may take has been highlighted by the cases in this book. To implement concurrent engineering effectively, companies must respond to this variety by configuring new product introduction differently. Various ways of constructing concurrent engineering solutions can be achieved by using different alternatives for each of the five elements of structure, process, control, people and tools. These form the basic building blocks of a product introduction configuration.

In order to build effective concurrent engineering solutions, it is not only necessary to have an idea of the options available, but also to have an understanding of the factors that will make one solution more effective than another. This is why it is useful to introduce the idea of forces or pressures which pull concurrent engineering solutions into an appropriate form for their environment. One way of considering these forces is the pentagon of forces of Focus, Incremental Change, Radical Innovation, Proficiency and Efficiency developed in Chapter 2. What the framework provides is a way of thinking through the issues that affect the selection of appropriate elements of a concurrent engineering solution that will actually fit the environment in which it must work. Whilst the framework will not on its own answer the question of which concurrent solution works where, it will certainly provide a tool to assist in that process.

Index

Senior management support 8–9, 160,
181, 182, 207–8, 228, 240
Sequential design processes 4, 8, 67, 96
Simultaneous engineering *see*
Concurrent engineering
Skills (of staff) 12, 25–6, 29, 36, 42,
63–4, 84, 88, 106, 112, 118,
214–216, 219–220, 225, 226, 227,
234, 237, 238, 240
see also Core competences; Education
of staff; Training of staff
'Skunkwork' teams (multidisciplinary
3M product development teams)
10
Specifications for products 36–7, 59–60,
74, 81, 98, 133, 212
Standards 65, 66, 105, 202
Structure in product introduction 45,
47–8, 49, 53–88, 242
Superteam Solution, The (book) 10
Supplier relationships 173–4, 183, 188,
200–203
Supply chains 18–19, 20, 202

Taguchi statistical technique 141, 179
Task forces *see* Teamwork (in projects)
Team based organizational structures
10
see also Multidisciplinary teams
Teamwork (in projects) 2, 3, 6, 12, 17,
38, 48, 66, 72–3, 74–6, 98, 99–102,
106, 197–8, 126–140, 208–9,
211–229, 231, 238
see also Co-location; Multidisciplinary
teams; Project planning and
management
Technology transfer 13
in projects 12

Temco Ltd 145–8, 163–184, 236
communication 171–184
concurrent engineering 145–8,
163–184, 236
multidisciplinary teams (in product
development) 171–184
organizational structure 171–3,
175–7, 183–4
product development and
introduction 145–148, 163–184,
236
Testing of new products 61, 132–140,
177, 197, 205–6
see also Problem monitoring
Time to market *see* Lead-times for new
products
Tools for product introduction 37, 38,
40, 45, 48–9, 80, 177–8, 242
see also Computerized tools
Total quality (concept) 8, 48
see also Quality management
Training of staff 147–8, 152, 158–9, 189,
207, 214–215, 219, 226–7
see also Education of staff

Uncertainty 31

Validation of data 108, 118, 123, 124
Variety in companies 26–31
Viable system model (VSM) 220–224
Videoconferencing 48, 119
Virtual teams 23, 89, 92, 101–2, 108, 119
'Voice of the Customer' (initiative in
Instron Ltd) 102–5
'Voice-planning' tables 40–41

Wire manufacturing industry 163–184

Loving Work

Frank Price

From the author of bestsellers *Right First Time* and *Right Every Time* comes a caustic but compassionate review of the working world of the mid-nineties.

Taking up his acclaimed theme once more, the author asks 'Where does quality go from here?' To answer this he looks at some of the options open to those with the drive to enrich their lives by enlarging their experience, whether they still have jobs or are working on their own. Its aim is to instruct, to entertain - above all to encourage.

Writing in the provocative and often hilarious style that his thousands of readers have come to relish, Frank Price uses jokes, military history, parable, anecdote and reminiscence to convey his message of strength, hope and ultimately love. Pulling no punches and leaving few sacred cows unscathed, he explores aspects of working life largely ignored by commentators, including bullying, fear and harassment. *Loving Work* is not a book for the fainthearted but it offers precious insights that can help us respond positively to the changes that confront us.

1995   255 pages   0 566 07634 9

Gower

Right Every Time
Using the Deming Approach

Frank Price

Over the five years since the publication of Frank Price's book *Right First Time* the business landscape of the Western World has undergone an upheaval - a Quality Revolution. This explosion of interest in the management of quality has not just affected the manufacturing sector but has influenced all areas of industry; and with diverse effects.

In *Right Every Time* the author not only examines the content of quality thinking, the statistical tools and their application to business processes; he also explores the context, the cultural climate, in which these tools are put to work, the environment in which they either succeed or fail. The core of the book consists of a critique of Deming's points - which the author refers to as the new religion of quality - and an examination of the pitfalls which act as constraints on quality achievement.

This is more than a 'how to do' book, it is as much concerned with 'how to understand what you are doing', and the book's message is applicable to anybody engaged in providing goods or services into markets where 'quality' is vital to business success. There can be no doubts concerning the benefits of quality control, and in this important and highly readable text Frank Price reveals how such visions of excellence may be transformed into manufacturing realities.

1990   192 pages   Hardback 0 566 09002 3   Paperback 0 566 07419 2

Gower

Right First Time
Using Quality Control for Profit

Frank Price

This remarkable book combines simplicity of treatment with depth of coverage and is written in a refreshingly original style. Dispelling the mystique which so often surrounds the subject, and without indulging in complex mathematics, the author explains how to achieve low scrap rates, zero customer rejections and the many other benefits of systematic quality control.

The twin themes of the book are the need for quality to be an integral part of the manufacturing process and the importance of commitment throughout the workforce. Thus it deals not only with QC concepts and techniques but also with the human and corporate relationships whose effects can be critical.

1984   320 pages   Hardback   0 566 02467 5   Paperback   0 7045 0522 3

Gower